D0893429

Indoor
Air
Research
Series

Max
Eisenberg
Series
Editor

Center
for
Indoor
Air
Research

RISK ASSESSMENT and INDOOR AIR QUALITY

Edited by
Elizabeth L. Anderson
and
Roy E. Albert

Lewis Publishers
Boca Raton   London   New York   Washington, D.C.

Library of Congress Cataloging-in-Publication Data

Risk assessment and indoor air quality / edited by Elizabeth L.
    Anderson and Roy E. Albert.
            p.   cm. -- (Indoor air research series)
        Includes bibliographical references and index.
        ISBN 1-56670-323-9 (alk. paper)
        1. Health risk assessment.   2. Indoor air pollution--Health
    aspects.   3. Ventilation--Health aspects.   4. Air quality--
    Evaluation.   5. Environmental risk assessment.   I. Anderson,
    Elizabeth L., Ph.D.   II. Albert, Roy E.   III. Series.
    RA566.27.R573   1998
    613'.5--dc21   98-26281
                                                                               CIP

Series Preface

The field of indoor air science is of growing interest and concern given that modern society spends the better part of each day indoors. Since the indoor air environment is a major, continual exposure medium for occupants, it is important to study what is present and if and how it affects the health and comfort of occupants.

Volumes in this Indoor Air Research Series are intended to provide state-of-the-art information on many areas germane to indoor air science including chemical and biological sources, exposure assessment, dosimetry, engineering controls, and perception of indoor air quality. In each volume, authors known for their expertise on the topic will present comprehensive and critical accounts of our current understanding in the area.

It is hoped that the series will advance knowledge and broaden interest among the scientific community at large in the indoor air science field.

Max Eisenberg, Ph.D.
Series Editor

Preface

Indoor air pollution was rarely identified as an environmental concern prior to the early 1970s. Since that time, however, both real and perceived indoor air problems have increased almost continuously. One reason is that buildings were tightened and air exchanges reduced to conserve energy. Another is that federal efforts began to control outdoor air pollutants, many of which also are indoor air pollutants. Finally, scientific techniques and methods began to be developed that provided better opportunities for quantifying the contaminants and their likely effects. One of the most important emerging tools is the science of risk assessment.

This book was commissioned by the Center for Indoor Air Research as a state-of-the-art review of the science of risk assessment and its application in understanding and remediating indoor air quality concerns. While the science of risk assessment and its uses for indoor air quality are well characterized and in growing use, both topics are rapidly evolving due to scientific, regulatory, political, and public concerns. Thus, this book was written to characterize the subjects, but at the same time to provide the necessary reference resources for more in-depth, future investigation. At the same time, it was written for use by readers with a wide range of educational and professional qualifications. It is the hope of the authors that the book will serve as a useful reference tool for advances and innovative solutions in these fields.

Elizabeth L. Anderson, Ph.D.
Roy E. Albert, M.D.
Editors

Acknowledgments

The authors would like to acknowledge Dr. Max Eisenberg and Dr. Lynn Kosak-Channing of the Center for Indoor Air Research for their support, insights, and patience during the preparation of this book.

The Editors

Elizabeth L. Anderson, Ph.D., is President and CEO of Sciences International, Inc. in Alexandria, Virginia and has over 20 years experience in risk assessment.

At the U.S. Environmental Protection Agency (EPA), she established and directed the central risk assessment program for ten years. She founded the Carcinogen Assessment Group and later was Director of the Office of Health and Environmental Assessment (OHEA) in the Office of Research and Development, with a staff of over 140 and an annual budget exceeding $14 million. The primary functions of OHEA were to provide leadership to establish EPA-wide guidelines for risk assessment, to conduct risk assessments on the health effects of toxic chemicals, and to oversee the risk assessment program for all of EPA's regulatory programs.

Since leaving EPA, Dr. Anderson has been engaged in managing governmental and private sector health and environmental consulting activities. She is an internationally recognized expert and lecturer and has published numerous journal articles in the areas of risk assessment and carcinogenicity. She was the recipient of the EPA Gold Medal for Exceptional Service and the Distinguished Service Award from the Society for Risk Analysis. She is a member of the American Association for the Advancement of Science and the New York Academy of Sciences. Her professional activities relating to risk assessment include the following:

Board of Scientific Counselors, Committee to Review the National Health and Environmental Effects Research Laboratory, U.S. Environmental Protection Agency, 1997.

Peer Review Committee, Exploratory Research Program, Environmental Physics, U.S. Environmental Protection Agency, 1997.

Peer Review Committee, Exploratory Research Program, Environmental Chemistry, U.S. Environmental Protection Agency, 1997 (reappointed for 1998).

Department of Defense Peer Review Committee, Strategic Environmental Research and Development Program (SERDP), 1997.

Chaired Peer Review Committee, Risk assessment guidelines for combustion sources, U.S. Environmental Protection Agency, 1996.

Peer Review Committee, Center for Risk Assessment, U.S. Environmental Protection Agency, 1996.

External Advisory Board, Center for Risk Management of Engineering Systems, University of Virginia, 1987–present.

Editorial Board for the journal *Human and Ecological Risk Assessment;* appointed 1994–present.

New York Power Commission Advisory Panel to recommend research programs to evaluate risk associated with electric and magnetic fields, 1990.

Risk Assessment Review Panel for the State of New Jersey; appointed 1988.

Panel of experts evaluating risk analysis activities of four federal agencies, General Accounting Office, for House Committee on Science and Technology, February 1986.

Charter Member, Society for Risk Analysis (member of steering committee to establish society, 1980); member of editorial board, Risk Analysis; elected council member, 1981; president, 1984–1985.

Subcommittee on Risk Analysis, Health and Environmental Research Advisory Committee, Department of Energy, 1985.

EPA Representative to the National Cancer Advisory Board, 1982–1985.

Interagency Risk Management Council, cabinet council committee; chairman, committee to develop guidelines for assessing reproductive risk.

International Program for Chemical Safety (IPCS) Committee Editorial Staff, principles for evaluating health risks to progeny associated with exposure to chemicals during pregnancy, World Health Organization, Geneva, Switzerland, 1984.

Interagency Regulatory Liaison Group, Work Group on Risk Assessment. (Work group published the article "Scientific Bases for Identification of Potential Carcinogens and Estimation of Risks," *JNCI* 63:242, 1979); Chairman of the work group, 1980.

Risk Analysis Liaison Committee, National Academy of Sciences/National Science Foundation (under P.L. 96–44).

National Academy of Sciences/Food and Drug Administration, Advisory Committee on institutional means for assessment of risk to public health (under H.R. 7591).

Roy E. Albert, M.D. specializes in research related to the quantitative aspects of chemical and radiation carcinogenesis. Dr. Albert has served as a consultant to various governmental and state committees, including: Surgeon's General Advisory Committee on Smoking and Health; Air Pollution Advisory Committee (New York City Department of Health); Ad Hoc Committee on Environmental Health Research — Panel on Hazardous Trace Substances for the Office of Science and Technology, Executive Office of the President; Motor Vehicle Nitrogen Oxide Standard Committee; and Committee on Water Treatment Chemicals for the National Research Council.

Dr. Albert is currently Professor of Environmental Health and Chairman of the Department of Environmental Health and the Kettering Laboratory at the University of Cincinnati Medical Center. Dr. Albert was the principle author of EPA's first carcinogen risk assessment guidelines and subsequently served for ten years as Chairman of the Carcinogen Assessment Group at the Environmental Protection Agency.

Dr. Albert received a Distinguished Contribution Award (Society for Risk Analysis, 1984). His professional affiliations include:

American Association for the Advancement of Science
American Association for Cancer Research
American College of Toxicology
New York Academy of Sciences
Radiation Research
Society for Epidemiological Research
Society for Occupational and Environmental Health

Contributors

Roy E. Albert, M.D.
Professor and Chairman
Department of Environmental Health
Kettering Laboratory (ML 56)
University of Cincinnati Medical Center
3223 Eden Avenue
Cincinnati, OH 45267-0056

Elizabeth L. Anderson, Ph.D.
President and CEO
Sciences International, Inc.
Suite 500
1800 Diagonal Road
Alexandria, VA 22314

Nicholas J. Gudka. M.S.
Program Manager
Sciences International, Inc.
Suite 500
1800 Diagonal Road
Alexandria, VA 22314

John J. Liccione, Ph.D.
Project Manager
Sciences International, Inc.
Suite 500
1800 Diagonal Road
Alexandria, VA 22314

Robert M. Little
Analyst
Sciences International, Inc.
Suite 500
1800 Diagonal Road
Alexandria, VA 22314

Suresh H. Moolgavkar, M.D., Ph.D.
Director, Moolgavkar Consulting
  Group
Sciences International, Inc.
9005 NE 21st Place
Bellevue, WA 98004

David R. Patrick, P.E.
Vice President
Sciences International, Inc.
Suite 500
1800 Diagonal Road
Alexandria, VA 22314

Steave H. Su, M.P.H.
Senior Associate
Sciences International, Inc.
Suite 500
1800 Diagonal Road
Alexandria, VA 22314

Lance A. Wallace
11568 Woodhollow Court
Reston, VA 20191
Affiliated with:
U.S. Environmental Protection Agency
Office of Research and Development
Research Triangle Park, NC 27711

Table of Contents

List of Tables

List of Figures

CHAPTER 1

Introduction to Risk Assessment

Elizabeth L. Anderson and David R. Patrick

CONTENTS

1-56670-323-9/99/$0.00+$.50
© 1999 by CRC Press LLC

I. OVERVIEW

Environmental risks can result from contact with a toxic material or contaminant via the environment. A human health risk can be experienced by individuals or populations from contact with an environmental contaminant through inhalation, ingestion, or skin contact. Such risks can be acute (short-term) or chronic (long-term) in nature and can range from mildly irritating to life threatening. Environmental risks also include ecological risks, such as effects on plants, animals, and natural resources, resulting from the presence of undesirable materials in the environment. Welfare risks are a type of environmental risk generally associated with the quality of human life (e.g., visibility, soiling, and weathering). In the indoor environment, human health risks are the principal concern. As such, this book focuses on human health risks principally resulting from indoor air exposures.

This book was prepared to provide guidance for identification of human health risks associated with indoor air exposures, estimation of the possible extent and severity of these risks, and determination of the effects of mitigation on these risks. This book is intended as a desk reference to assist readers in making more informed decisions regarding the need and appropriate means for improving indoor air quality. Indoor air quality decisions that can benefit from the use of risk assessment include the following:

- setting priorities for study or mitigation of risks resulting from indoor air quality,
- determining proper avenues of evaluation or investigation of these risks,
- establishing criteria for the timing and degree of mitigation of these risks,
- identifying and selecting appropriate mitigation strategies,
- identifying appropriate research needs, and
- assisting in regulatory decision-making.

This chapter presents a brief history of the origin and development of risk assessment as well as an introduction to its application in indoor air quality studies. Chapter 2 defines the risk assessment process and describes its origins both scientifically and legislatively; Chapters 3 through 6 provide detailed discussions of the four principal components of a risk assessment; Chapter 7 discusses the uncertainties associated with risk assessment; Chapter 8 describes basic methods for measuring indoor air contaminants; Chapter 9 presents a case study of the application of risk assessment to a typical indoor air problem; and Chapter 10 identifies future risk assessment directions and needs.

II. WHAT IS RISK ASSESSMENT?

Risk is generally defined as the potential for an unwanted negative consequence or event. As used in this book, *risk* is limited to unwanted adverse human health effects resulting from exposures in the indoor environment. *Risk* should be distinguished from *hazard*. A hazard is a possible source of danger; however, a risk is not present unless a human can come into contact with or be exposed to the hazard. A risk assessment in this context is the systematic evaluation of the factors that might result in an adverse human health effect resulting from a hazard, and often the attempted quantification of those factors and effects. As described by the National Research Council (NRC 1983), risks are assessed for a variety of reasons, one of the most important of which is regulatory decision-making. The results of the risk assessment are dealt with in a process usually called risk management. The distinctions between risk assessment and risk management are discussed more fully in later sections.

Ideally, risk assessment is based in science and risk management is the policy for use of that science. In reality, however, the distinctions are often not so clear. For example, policy choices are often required in the risk assessment and these can often significantly affect its outcome. In addition, the effects of exposure by animals and humans to toxic substances are not always well understood by scientists, often because the organisms and the interactions are so complex, or because the effects can vary within and across species. As such, assumptions must be made to allow scientists to extrapolate results across species and across ranges of exposure. Policy choices can influence the selection of these assumptions. A conservative (i.e., health protective) safety factor may be selected rather than a more moderate safety factor to minimize the unwanted consequences of error. What this means is that policy choices are intertwined with scientific determinations. Another difficulty in the risk management process is that regulatory decision-makers dislike uncertainty because it complicates the decision-making process, often forcing the use of conservative assumptions that may be economically undesirable. Early attempts at risk assessment and risk management aimed at providing specific health criteria, including workplace limits and national ambient air quality standards. Currently, attempts have been made to provide a broader risk assessment/risk management framework for decision-making that may include a variety of information such as exposure distributions, ranges of health effects, and even economic consequences of regulatory actions.

It is important that the reader recognize that risk assessment will rarely provide complete and unequivocal results for decision-making. The science of risk assessment is still in its relative infancy and it is complex. As such, risk assessment is, and will continue to be, associated with uncertainty. Typical areas of uncertainty with respect to air quality (indoor or outdoor) risk assessments include the following:

- large variations in measured data and in human responses to environmental exposures;
- limited understanding of the toxicology and exposure pathways for many contaminants;
- improperly designed or understood mathematical models;

- the unique nature of individual human exposures to the array of contaminants in his or her environment;
- imprecise knowledge of the contaminants to which humans are exposed; and
- the vast variety of possible exposures.

Still, enough is known in many instances today to allow risk assessment to be used as a tool with growing application and precision in decision-making. This book is intended to guide a reader with responsibilities or concerns about indoor air quality in identifying important air quality and health issues and in conducting analyses sufficient to facilitate responsible decision-making. It also is written for the reader who is technically experienced although not necessarily in the science of risk assessment.

III. INDOOR AIR RISK ASSESSMENTS

The term *indoor environment*, as used here, encompasses all enclosed spaces occupied by humans, including home, work, shopping, education, entertainment, and transportation. While humans can be exposed indoors to contaminants by inhalation, ingestion, and dermal contact, the inhalation pathway usually dominates indoor air quality investigations, and thus this book focuses on human health risks resulting from inhalation. However, risks from other pathways should be considered if there is information or strong evidence that another pathway can contribute significantly to a potential adverse human health effect. One example might be a biological contaminant that can be conveyed through inhalation and skin contact; another example is a contaminant found in the air of a household and, concurrently, in food eaten by members of the household.

Humans can be exposed to environmental risks outdoors or indoors. However, since we first began to control pollutants in the air that could adversely affect humans or the quality of life, most attention has focused on air pollutants in the outdoors and assumed outdoor exposures. Researchers now recognize, however, that most of the population spends the bulk of its time indoors and that indoor exposures are more important than, or at least as great a concern as, outdoor exposures. There are a number of reasons why the types and concentrations of indoor air pollutants are growing. For example, the energy crisis beginning in the early 1970s led architects, engineers, builders, building managers, and home owners to take steps to conserve energy, including reduction in the infiltration of outside air, recirculation of building air, and greater use of synthetic building and decorative materials. While these actions generally achieved their purpose of reducing energy costs, they often resulted in increasing indoor concentrations of chemical and biological substances arising from both indoor and outdoor sources. In addition, the synthetic materials and decorations increasingly being used in homes and buildings can release new chemicals into the indoor environment. Although debate continues concerning the causes, many scientists believe that these buildups in indoor air concentrations coincided with a growing increase in indoor air quality related illnesses of both specified and unspecified natures.

In its simplest form, an assessment of possible indoor air risks leads to the determination of an acceptable exposure limit for specific substances to which a human can be exposed. These exposure limits usually are derived by expert scientific judgment or through the application of accepted safety factors to animal test results. Acceptability is determined by comparing actual exposures with an accepted limit. If humans are exposed to concentrations less than the limit, then the exposure usually is judged acceptable; if the exposures are greater than the limit, then guidance usually specifies that the humans should be removed from the exposure or the exposures should be otherwise reduced. Acceptable workplace limits for air pollutants are published by numerous regulatory and quasi-regulatory bodies both in the U.S. and abroad. In the U.S., these include the Occupational Safety and Health Administration (OSHA), state and local agencies, and the American Conference of Governmental Industrial Hygienists (ACGIH). Internationally, the World Health Organization (WHO) plays a leading role in Europe, and individual European countries, Canada, and Japan have active air pollutant regulatory programs. Most of these organizations recognize the requirement for expanded indoor air quality programs.

Unfortunately, the process of setting acceptable exposure limits begins to break down when the adverse effects resulting from exposures are not adequately represented by a simple pass-fail test. This first became apparent when U.S. regulators attempted to regulate carcinogens in the 1970s. Most suspect carcinogens do not have an identifiable, lower threshold of effect. This factor was interpreted as meaning that any exposure is associated with a risk and that regulators must decide what level of risk is "acceptable." Many people argued that no man-made risk is acceptable and that man-made sources of cancer risk should be eliminated; others recognized that the elimination of man-made sources of cancer risk would have serious economic consequences. Regulators were faced with a conundrum epitomized by lapel pins at several public meetings in 1983 in Tacoma, Washington, the site of the largest source in the U.S. of inorganic arsenic, a human carcinogen. The pins stated simply "Jobs or Lives." Fortunately, federal, state, and local regulatory officials were able to defuse the passions of the moment and went on to implement regulatory controls that did not immediately shut down the plant (although it did later close for a variety of reasons) and were convincing enough that the community accepted them with new pins stating "Jobs and Lives."

Regulation of indoor air exposures is difficult for other reasons. For example, for some time investigators have known that occupants in some buildings exhibit health symptoms including eye, ear, nose and skin irritation, dry mucous membranes and skin, respiratory infections and cough, hoarseness of voice and wheezing, hypersensitivity reactions, nausea and dizziness, and mental fatigue and headache that appear to be relieved when they leave the building. These symptoms occur frequently enough that they have come to be known as Sick Building Syndrome. Rarely can the symptoms be traced to a specific substance or action, and while many investigators believe that the effects are real, others believe the syndrome is in large part due to psychological factors such as job stress. A similar controversy is whether some individuals are hypersensitive to very low concentrations of chemical mixtures. This, too, occurs frequently enough that it has come to be known as Multiple Chemical Sensitivity. Again, adverse effects have not been traced to specific mixtures or concentrations and, in the individuals apparently affected, there are differences in response, sensitization, desensitization, and other biological factors. In both cases, research remains to be conducted both to understand the underlying causes and to develop appropriate solutions where the effects are shown to be valid.

The confidence in a given acceptable indoor or outdoor exposure limit is also a function of the confidence in understanding the potential health effects associated with exposures, which may come from human and animal studies, and how the test exposures are extrapolated to real-world exposures. Uncertainty is often dealt with by applying safety factors or by assuming worst-case exposures. No matter how it is represented, uncertainty is almost always dealt with by making conservative assumptions. This bias has been adopted because public health officials must make decisions in the face of scientific uncertainty. If there is error, the choice is to err on the side of public health protection. The potentially high economic and social costs of some "conservative" decisions argue strongly for developing more and better data to reduce uncertainty. On one hand, the higher quality data frequently results in health limits perceived to be less restrictive because there is reduced need for conservative assumptions. On the other hand, the potential costs also often lead to the development of more precise decision tools to facilitate more appropriate and informed decisions.

IV. IMPORTANT INDOOR AIR AND RISK ASSESSMENT DEFINITIONS

Absorbed dose — The amount of an agent that enters the body (see *Internal dose*).

Acceptable risk — A level of risk that is considered low enough to be deemed insignificant or *de minimis*. For example, the EPA established cancer risk criteria for benzene in 1989 that requires protection of the greatest number of people to an individual lifetime cancer risk no greater than 1 in 1,000,000 and limiting to no greater than 1 in 10,000 the individual lifetime cancer risk of the most exposed individual. In California's product labeling law, an incremental lifetime risk of 1 in 100,000 is considered insignificant.

Accuracy — The measure of the correctness of data, as given by the difference between the measured value and the true or standard value.

Acute effect — Occurring over a relatively short period of time, particularly an adverse health effect that appears promptly after exposure.

Acute exposure — A relatively short-term exposure; the OSHA often establishes acute workplace exposure limits for 15-min exposures and ceiling (i.e., peak) exposures. The EPA also establishes outdoor air standards for acute exposures, usually one hour.

Agent — A chemical, physical, mineralogical, or biological entity that may cause a deleterious effect in an organism after exposure; also called a contaminant or pollutant.

Ambient — Generally the outdoor environment or surrounding conditions.

Antagonism — Interference or inhibition of the effect of one agent by the action of another.

Applied dose — The amount of a substance in contact with the primary absorption boundaries of the organism (e.g., lung, skin, and gastrointestinal tract) and available for absorption.

Background level — Normal environmental concentrations of an agent before introduction of new quantities through emission or release.

Bias — A systematic error inherent in a method or caused by some feature of the measurement system.

Bioaccumulation — The retention or concentration of a substance in a media or organism.

Biological marker or biomarker — An indicator of changes or events in human biological systems, generally referring to cellular, biochemical, or molecular measures obtained from human tissue, cells, or fluids and indicative of exposure to an environmental contaminant.

Biologically effective dose — The amount of the deposited or absorbed contaminant that reaches the cells or target site where an adverse effect occurs or where an interaction of that contaminant with a membrane surface occurs.

Breathing zone — The air in the vicinity of an organism from which respired air is drawn. Personal monitors often are used to measure pollutants in the breathing zone.

Carcinogen — A substance that can cause or induce cancer in humans or animals.

Cancer potency factor — A numerical factor expressed as the reciprocal of dose and representing the strength of a carcinogen; at a unit dose, the term is called the unit risk factor. Multiplying the cancer potency factor by the dose provides a numerical probability of getting cancer.

Chronic effect — Occurring over a relatively long period of time, particularly an adverse health effect that appears after long-term, low-level exposures.

Chronic exposure — A relatively long-term exposure; the OSHA often establishes a chronic workplace exposure limit for 8-hour work day and 40-hour work week exposures. The EPA also often establishes outdoor air standards for chronic exposures, including daily, monthly, and annually.

Concentration — The accumulation of an agent in plants, organisms, or other receptors to levels generally greater than the level in the media resulting in the exposure.

Degradation — Chemical or biological decomposition of a substance into elementary substances.

Delivered dose — The amount of the contaminant that is transported to the organ, tissue, or fluid of interest.

Demography — The study of the characteristics of the human population, including size, growth, density, distribution, and vital statistics.

Dermal exposure — Contact between an agent and the skin.

Dose — The amount of a contaminant that is absorbed or deposited in the body for an increment of time, usually from a single medium. Total dose is the sum of doses received by the person from all environmental media that contain the contaminant.

Dose–response — A quantitative relationship between the dose of an agent and an effect caused by the agent.

Dose–response assessment — The determination of the relationship between the magnitude of the applied or internal dose and a specific biological response.

Environment —The air, water, surfaces, and food to which a person is exposed; generally includes all indoor and outdoor environments.

Environmental fate — The destiny of an agent after release to the environment. It can involve consideration of transport through the air, soil, and water, as well as concentration, degradation, and other factors.

Epidemiological studies — The investigation of human populations to assess the incidence and possible causes of disease.

Exposure — Contact with a chemical, physical, or biological agent at the outer boundary of the organism. Exposure is quantified as the concentration of the agent in the medium of contact integrated over the duration of the contact.

Exposure assessment — The determination or estimation (qualitative or quantitative) of the magnitude, frequency, duration, route, and extent (i.e., number of people) of exposure to an agent.

Exposure concentration — The concentration of an agent at the point of contact.

Exposure pathway — The route taken by an agent as it travels from its source to a receptor.

Exposure route — The way an agent enters an organism after contact (e.g., by inhalation, ingestion, or dermal absorption).

Exposure scenario — A set of conditions or assumptions about sources, exposure pathways, concentrations of agents, and populations (i.e., numbers, characteristics, and habits) that aid in the evaluation and quantification of exposure in a given situation.

Extrapolation — Estimation of unknown values by extending or projecting from known values.

Hazard — In this context, a substance associated with an inherent ability to result in an adverse health effect in humans if the human inhales, ingests, or comes in contact with the substance. A hazard is distinguished from a risk that describes the type and severity of the adverse effect after exposure.

Hazard identification — A description of the potential health effects attributable to a specific chemical, physical, or biological agent. For carcinogen assessments, the hazard identification step is also used to determine whether the particular agent is, or is not, causally linked to cancer in humans.

High-end exposure (dose) estimate — As used by the EPA, a plausible estimate of population exposure or dose for those persons at the upper end of an exposure or dose distribution, conceptually above the 90th percentile, but not higher than the individual in the population who has the highest exposure or dose.

High-end risk descriptor — A plausible estimate of the individual risk for those persons at the upper end of an exposure or dose distribution, conceptually above the 90th percentile, but not higher than the individual in the population with the highest risk. Since high risk may result from high exposure, high susceptibility, or other reasons, the persons in the high-end of the exposure distribution may not be the same persons in the high-end of the risk distribution.

Indoor risk assessment — An assessment that covers a broad range of potential health concerns, including radon, biological agents, environmental tobacco smoke, outdoor ambient pollutants, and a wide variety of pollutants in the indoor environment.

Intake — The process by which a substance crosses the outer boundary of an organism without passing an absorption barrier. (See *Potential dose.*)

Integrated exposure assessment — An integration of all relevant information and summation over time of the magnitude of exposure to an agent.

Internal dose — The amount of a substance penetrating across the absorption barriers of an organism, through either physical or biological processes; generally synonymous with absorbed dose.

Maximally (or most) exposed individual (MEI) — The single individual with the highest exposure in a given population. Historically, this term has been defined in various ways, including worst case exposure.

Meteorology — The weather patterns and characteristics that influence the movement and dispersion of air pollutants from their sources.

Microenvironment — A three-dimensional space in which the concentration of an agent or agents is uniform during a specified interval; includes the home, office, automobile, kitchen, shopping, and all other locations that can be well-characterized in concentrations of an agent.

Modeling — Use of mathematical relationships to simulate and predict real events and processes.

Monte Carlo analysis — A repeated random sampling from the distribution of values for each of the parameters in an exposure or dose equation to derive an estimate of the distribution of exposure or dose in a population.

Multipathway — Involving consideration of all pathways through which exposure occurs. The three primary human exposure pathways are inhalation, ingestion, and skin contact.

Nuisance effect — A subjectively unpleasant effect (e.g., headache) that occurs as a consequence of exposure to a contaminant. These effects are not permanent.

Pathway — The physical course an agent takes from its source to the exposed organism.

Potential dose — The amount of an agent contained in material ingested, air breathed, or material applied to the skin.

Precision — A measure of the reproducibility of a measured value under a given set of conditions.

Qualitative — Descriptive of kind, type, or direction.

Quantitative — Descriptive of size, magnitude, or degree.

Reasonable worst case — As used by the EPA, a semiquantitative term referring to the lower portion of the high end of the exposure, dose, or risk distribution. Historically, this term has been loosely defined, often considered synonymous with maximum exposure or worst case. (See also *High-end exposure estimate.*)

Receptor — In exposure assessment, the organism that receives, may receive, or has received environmental exposure to a contaminant.

Reference concentration (RfC) — For noncarcinogens, the estimate of the concentration of a substance that is likely to be without appreciable risk of deleterious effect during a lifetime of exposure to a person; often used when inhalation is the principal route of exposure.

Reference dose (RfD) — For noncarcinogens, the estimate of the daily dosage to a substance that is likely to be without appreciable risk of deleterious effect during a lifetime of exposure to a person; often used when ingestion or skin contact is the principal route of exposure.

Representativeness — The degree to which a sample is, or samples are, characteristic of the whole medium, exposure, or dose for which the samples are being used to make inferences.

Risk — The probability that a specific unwanted health effect may occur as a result of a specified exposure to an agent.

Risk assessment — A qualitative or quantitative evaluation of the health or environmental risk resulting from exposure to an agent. A risk assessment combines the results of the exposure assessment and the toxicity assessment to estimate risk.

Risk characterization — The description of the nature and often the magnitude of human or nonhuman risk, including the attendant uncertainties.

Route of exposure — The avenues by which an agent comes into contact with an organism, usually though inhalation, ingestion, or skin contact.

Source characterization measurements — Measurements made to characterize the rate of release of agents into the environment from a source.

Topography — The physical features of an area. The extent of human exposure can be influenced by the presence of mountains, valleys, bodies of water, and other topographical features.

Total human exposure — Accounting for all exposures of a person to a specific contaminant from all media and through all routes of entry.

Toxic — The condition of being harmful, destructive, or deadly.

Toxicity — The quality or degree of being poisonous or harmful.

Toxicity assessment — Characterization of the toxicological properties and effects of an agent, including all aspects of its absorption, metabolism, excretion, and mechanisms of action.

Upper bound estimate of risk — As used by the EPA, a conservative estimate of risk made in the absence of specific information. The true risk, if it could be known, should almost always be lower than the upper bound estimate.

Uptake — The process by which a substance crosses an absorption barrier and is absorbed into the body.

Worst case — As used by the EPA, a semiquantitative term referring to the maximum possible exposure, dose, or risk that can conceivably occur, whether or not it actually occurs or is observed. This typically refers to a hypothetical situation in which everything that can plausibly happen to maximize exposure, dose, or risk, in fact, does happen. While it is conceivable that this worst case could occur in a given population, the worst case is almost always higher than occurs in a specific population. The worst case scenario is most valuable in evaluating low probability events that may result in a catastrophe that must be avoided even at great cost. In many health risk assessments, a worst case scenario serves as the upper bound.

V. THE ORIGINS OF ENVIRONMENTAL RISK ASSESSMENT

A. Environmental Risk Assessment Prior to 1970

Humans have always estimated the risks of their actions or inactions in making personal decisions. However, the process was either intuitive or empirical until the mid-17th century when probabilities began to be described mathematically, initially to calculate gambling odds more precisely and later to calculate the odds of life events, such as the expected age of death for life insurance policies. Environmental risk was not assessed quantitatively on a broad scale until the advent of nuclear power when public concerns arose over the potentially disastrous and long-term effects of nuclear accidents. These risk assessments were among the first that estimated both the likelihood of an undesirable occurrence and the magnitude of the impact on humans and the environment.

Congress and other regulatory bodies generally ignored environmental risk until Congress addressed risk qualitatively in the Delaney Clause of the Food Additive Amendments of 1958. This clause stipulated that no additive found to cause cancer in humans or animals could be allowed in the food supply. The policies that resulted from that clause led to prohibition of exposures to some substances believed to be carcinogens — a zero risk-tolerance policy. While commendable in its public health intent, the clause encourages uninformed decisions because it does not allow for consideration of the uncertainty of hazard, the magnitude of risk, or the concurrent benefits of the additives. For example, the addition of saccharine as a sweetener in food was initially banned although the benefits of a nonnutritive sweetener to diabetics and dieters are believed by many to outweigh the very low estimated cancer risks that might result from consumption of the added saccharine.

Before the EPA was formed in 1970, the responsibility for regulating the environment rested largely in the hands of the states. Their responses to environmental issues varied widely and were generally directed at highly visible problems such as air pollution from Pittsburgh's steel industry, smog resulting from Southern California's rapidly growing automobile population, and air pollution related deaths in 1948 in Donora, Pennsylvania. The formation of the EPA was based in large part on the growing conviction that a stronger federal oversight and abatement authority was necessary to ensure equal protection to all citizens and to address the growing interstate nature of air pollution and its sources.

B. The Use of Risk Assessment in the U.S. for Regulating Air Pollutants

1. Early EPA Regulatory Efforts

The EPA initially concentrated on establishing concentration standards for exposures to air and water pollutants and on publishing control technology guidance. The work on air pollution was required by the passage of the 1970 Clean Air Act Amendments (PL 91-604, December 31, 1970). Two types of air pollutants were identified by Congress for regulation under the 1970 amendments:

> *Criteria air pollutants* — These are air pollutants that "endanger public health and welfare"[1] and result from "numerous or diverse mobile or stationary sources." The EPA was required to establish "criteria" (i.e., all identifiable effects) for these pollutants, publish national ambient air quality standards (NAAQS) that allow an "adequate margin of safety" to protect the public health, and control them in a joint program with the states.
>
> *Hazardous air pollutants* — These are pollutants that reasonably may be anticipated to result in an "increase in mortality or an increase in serious irreversible, or incapacitating reversible, illness." These pollutants were to be listed by the EPA and regulated to achieve an "ample margin of safety to protect the public health."[2]

The EPA quickly listed several criteria air pollutants and initiated the mandated programs that, with amendments, continue to deal with these pollutants. Today, six criteria air pollutants are regulated:

---

[1] Welfare effects include but are not limited to effects on soil, water, crops, vegetation, man-made materials, animals, wildlife, weather, visibility, and climate, damage to and deterioration of property, and hazards to transportation, as well as effects on economic values and on personal comfort and well-being.

[2] Congress left it to the EPA to define both adequate margin of safety and ample margin of safety.

1. Particulate matter[3]
2. Ozone[4]
3. Sulfur oxides
4. Nitrogen oxides
5. Carbon monoxide
6. Lead

The EPA was required to establish the NAAQS and to review the standards at least every five years in a scientifically and publicly reviewed process that over time became increasingly resource-intensive and time-consuming. This process seeks to establish a threshold health effects level that protects the public from an unacceptable risk of harm while considering the nature and severity of the effects, the sensitive populations, and the uncertainties involved. While largely produced and resident in the outdoors, all of the criteria air pollutants can infiltrate into, and several can be produced, indoors and can affect indoor populations. However, none of the current Clean Air Act criteria air pollutant control strategies directly address the indoor environment.

For hazardous air pollutants, Congress left it to the EPA to identify the candidates and develop appropriate control strategies. While enacted for outdoor air pollutants, the debate and controversies surrounding hazardous air pollutants are relevant to indoor air risk assessment. The EPA initially regulated asbestos, mercury, and beryllium. The EPA established safe ambient exposure levels for mercury (neurological damage) and beryllium (lung disease) and subsequently promulgated regulations for industrial sources of these pollutants (40 CFR Part 61, Subparts C, D, and E). However, the EPA was unable to establish a safe level for asbestos because asbestos exposure can cause cancer, and there was no means at that time for deciding how to regulate carcinogens to achieve the required "ample margin of safety." Neither were there any generally accepted methods for measuring asbestos emissions or exposures. Thus, the EPA promulgated regulations that required "no visible emissions" from various asbestos sources (40 CFR Part 61, Subpart M). The EPA reasoned that "no visible emissions" represented an ample margin of safety; this was generally accepted because there was no reasonable alternative apparent.

The next hazardous pollutant that the EPA set out to regulate was vinyl chloride, which was also associated with cancer in workers exposed to vinyl chloride. Since there was still no guidance or method for regulating carcinogens with an ample margin of safety, the EPA promulgated standards establishing concentration limits on emissions of vinyl chloride (40 CFR Part 61, Subpart F). An environmental group quickly filed a lawsuit claiming that the standards should be more strict. This suit occurred at a time when public concern over environmentally caused cancer was

[3] Beginning in 1987, EPA limited regulation of particulate matter to particles less than or equal to 10 microns in diameter (called PM_{10}). On July 18, 1997 (63 FR 38702-38752), EPA promulgated additional standards for particles less than or equal to 2.5 microns in diameter (called $PM_{2.5}$).

[4] Ozone is rarely directly emitted to the air but is the component of most concern in photochemical smog. Photochemical smog is formed in a sunlight-catalyzed reaction between volatile organic compounds and nitrogen oxides in the air; thus, ozone is largely regulated by controls on the volatile organic compounds and nitrogen oxides.

growing rapidly, and many believed that carcinogens should not be allowed to be released (i.e., zero emissions of carcinogens) into the environment. An out-of-court settlement was reached on the vinyl chloride case requiring more restrictive concentration limits and the commitment by the EPA to pursue the zero-emissions solution.[5] However, concerns quickly resurfaced when the EPA soon thereafter listed benzene, a known workplace carcinogen, as a hazardous air pollutant (42 FR 29332, June 8, 1977).

As the hazardous air pollutant debate proceeded, the EPA was organizing to address the scientific questions of environmental carcinogens. Early focus was on several economically important pesticides including DDT, aldrin/dieldrin, and chlordane/heptachlor. The failure of a zero risk tolerance policy for suspect carcinogens led the agency to develop its first policies to guide the regulators through these issues. This work culminated in interim procedures and guidelines for assessing risk associated with exposure to suspected carcinogens in 1976 (41 FR 21402, May 25, 1976) and the proposal of an air cancer policy in late 1979 (44 FR 58642, October 10, 1979). The interim carcinogen procedures and guidelines for suspected carcinogens set forth a framework for scientifically determining the weight of evidence and magnitude of risk associated with suspect carcinogens; the air cancer policy proposed a framework for making regulatory decisions on carcinogens. Neither provided numerical targets and the air cancer policy was never finalized by the agency; however, adoption of risk assessment for evaluating suspect carcinogens was the landmark decision that initiated the risk assessment and risk management process for regulating environmental agents.

One reason for the lack of progress in managing risk under the Clean Air Act was that the EPA had no specific legislative guidance for regulating hazardous air pollutants. For example, there was no accepted definition of ample margin of safety. In addition, zero emissions are generally impossible to achieve without source closure, which could have significant economic impacts. Finally, there were literally tens of thousands of chemicals in commerce, many of which could be associated with adverse human health effects that might meet the definition of hazardous air pollutant but for which there was inadequate time and resources for evaluation. All of these issues led the agency to continued analysis but to little actual regulation.

In 1983, the General Accounting Office (GAO) released a report highly critical of the EPA's lack of action in regulating hazardous air pollutants. A Congressional hearing was held in 1983 on the GAO report, and a series of public hearings were held in the same year in the state of Washington on the regulation of inorganic arsenic, a carcinogen listed as a hazardous air pollutant in 1980 whose largest U.S. source was in Tacoma. The EPA's Administrator at the time, William Ruckelshaus, stated that he wanted input from those people potentially exposed to carcinogenic emissions before making a final decision on the degree of control. While these activities highlighted the issues and led to new strategies by the EPA and final regulation of arsenic in 1987, there still was no general resolution on how to regulate carcinogens and what comprised an ample margin of safety.

[5] The EPA never fully implemented the agreement, ultimately leading, as discussed below, to more litigation.

The long-running vinyl chloride debate finally reached the U.S. Court of Appeals in 1986 and the issues were settled. The Court ruled (Natural Resources Defense Council, Inc. v. EPA, 824 F.2d 1146[1987]) that the Clean Air Act did not require zero emissions for carcinogens and other environmental pollutants for which there is no safe threshold, but that practical factors such as cost and technical feasibility could not be used in setting standards. Instead, the Court set forth a two-step process for dealing with hazardous air pollutants:

1. The EPA must establish a safe or acceptable risk level which does not consider cost or technical feasibility. The Court stressed that this does not mean a risk-free level or that it had to be free of uncertainty. Rather, safe was defined as "acceptable in the world in which we live."
2. The EPA was to set standards that could be equal to or lower, but could not be higher, than the safe or acceptable level to protect the public with an ample margin of safety.

This ruling finally required the EPA to establish a safe or acceptable risk level for carcinogens. As noted earlier, the EPA proposed, but did not finalize, an Air Cancer Policy in 1979 (although it did not establish numerical targets). Finally, in late 1989, following public review, the EPA published a response to the Court's imperative as part of a regulatory decision on benzene (54 FR 38073, September 14, 1989). The EPA's approach for complying with Court's two-step process was to protect the public health with an ample margin of safety by providing maximum feasible protection against risks to health from hazardous air pollutants using the following two steps:

1. Protect the greatest number of persons possible to an individual lifetime risk level no higher than approximately 1×10^{-6},[6] and
2. Limit to no higher than approximately 1×10^{-4} the estimated risk that a person living near a source would have if he or she were exposed to the maximum pollutant concentrations for 70 years.

The EPA finalized benzene regulations using this approach but took no further regulatory actions before enactment of the 1990 Clean Air Act Amendments which dramatically changed the way hazardous air pollutants were to be regulated.

2. The 1990 Clean Air Act Amendments

In writing Title III (Hazardous Air Pollutants) of the 1990 Amendments, Congress apparently decided that the EPA was not moving quickly enough on these pollutants and that risk assessment was still controversial. Therefore, section 112 detailed the following requirements:

---

[6] The scientific notation 1×10^{-6} means one divided by one million; one million is one followed by six zeroes.

- Congress preempted the EPA's previous listing responsibility by specifically listing 189 substances[7] as hazardous air pollutants (HAPs). They required the EPA to set technology standards for all major sources[8] of these pollutants. The EPA subsequently listed 174 source categories that were believed to include all major sources of HAPs; Congress required that the EPA promulgate all technology standards by the year 2000.
- Congress required the National Academy of Sciences to submit a report to Congress by May 1993 that reviewed risk assessment methodologies used by the EPA to determine carcinogenic risk and to recommend improvements in that methodology. The required report actually was completed and submitted to Congress in 1994. It is discussed below.
- A new Risk Assessment and Risk Management Commission was formed by the 1990 Amendments, and the Commission was required to investigate the policy implications and appropriate uses of risk assessment in regulatory programs. The report of the Commission was published in early 1997.
- The EPA was required to report to Congress by November 1996 on methods for calculating risks, the public health significance of those risks, the actual health effects associated with exposures to HAPs, and recommendations for changes. At the time this book was written in mid-1997, the required report was in preparation and scheduled for release in draft form in early 1998.
- Section 112 provided that if Congress did not act on any recommendation provided by the EPA in the preceding report, the EPA is required, within eight years of promulgating a technology standard for a HAP source, to evaluate the risks remaining after application of the technology standards and to determine whether the public health is being protected with an ample margin of safety. If it is not, then more stringent standards are required. Congress further required that if the technology standards do not result in a reduction in lifetime excess cancer risks to the most exposed individual to less than 1×10^{-6}, the EPA is to promulgate additional standards. Importantly, Congress expressly reaffirmed the 1989 benzene risk assessment decision process.

3. Current Activities

At the time of preparation of this book, the EPA's hazardous air pollutant risk assessment program was awaiting completion of the required report to Congress on risk assessment and possible Congressional action on that report. The eventual policies or regulations, when complete, will legally apply only to outdoor hazardous air pollutants as required to be regulated by the 1990 Clean Air Act Amendments. However, the general lack of acceptable risk criteria and decision processes in other EPA programs and the lack of other specific legislative requirements for dealing with environmental risks means that the eventual outdoor hazardous air pollutant policies and procedures could be applied more broadly to the evaluation and reduction of human health risks related to indoor air.

---

[7] One substance was subsequently deleted; at the time of this writing in 1997, there were 188 listed HAPs.
[8] A major HAP source is one that emits more than 10 tons per year of a single HAP or 25 tons per year of any combination of HAPs.

C. Risk Assessment in the European Community and the United Kingdom

In late 1995, the law firm of Covington and Burling in London published a Memorandum entitled *Methods of Risk Assessment in the European Community and the United Kingdom* (Covington & Burling 1995). This memorandum provided information on risk assessment schemes used in Europe. It noted that risk assessment requirements generally begin at the European Community (EC) level but are implemented on the national level. As in the U.S., regulators set the standards and perform some risk assessments, but they often rely on industry to perform assessments with regulatory supervision.

A series of EC Council Directives requires employers to assess health and safety risks to their employees, including Council Directive 89/391, on the introduction of measures to encourage improvements in the safety and health of workers at work, and Council Directive 90/394, on the protection of workers from the risks related to exposure to carcinogens at work. While both require the determination and assessment of risks, there is scant information on the precise means and tools for conducting the risk assessments. Council Directive 80/1107/EEC, on the protection of workers from the risks related to exposure to chemical, physical, and biological agents at work, also provides only general requirements.

In the United Kingdom, the Health and Safety Executive (HSE) has issued several guidelines for risk assessment, including *Five Steps to Risk Assessment, Assess the Risks,* and *A Step by Step Guide to a Safer and Healthier Workplace.* The HSE has stated that it does not expect employers to undertake quantitative risk assessments. The techniques discussed in the guidelines include, for example, simplified methods where the probability, severity, and frequency of a risk are each assigned a factor on a scale (two scales used are 1 to 4 and 1 to 10) and the values are multiplied together, sometimes along with the number of people exposed, to determine a subjective risk level.

In the U.K., the Control of Substances Hazardous to Health (COSHH) Regulations of 1994 along with other guidance documents implement most of the requirements of the EC directives. Important documents include the Approved Codes of Practice (ACOP) series General COSHH ACOP, Carcinogen ACOP, and Biological Agents ACOP, and the Environmental Hygiene (EH) series document EH40/95, Occupational Exposure Limits. The COSHH ACOP regulations impose certain obligations on employers to assess the risks created by work. Again, the regulations do not provide detail on risk assessment methodologies but do require review and acceptance. The regulations do require specific actions in preventing or controlling exposure, using control measures, monitoring exposures, conducting health surveillance, and providing information, instruction, and training. In addition, exposure to specific chemicals and substances is limited, prohibited, or regulated in various ways.

The COSHH regulations are also the basis for the establishment and use of occupational exposure limits. Two types of limits are identified. The maximum exposure limit (MEL) is an exposure level at which a residual risk may exist; the level is set taking into account socioeconomic factors. The occupational exposure standard (OES) is set at a level at which there is no indication of a risk to health. The COSHH requires that where there is exposure to a substance for which an MEL is specified, the control of the exposure shall only be treated as being adequate if the level of exposure is reduced so far as is reasonably practicable and, in any case, below the MEL. Where there is exposure to a substance for which an OES has been approved, the control of exposure shall be treated as being adequate if the OES is not exceeded or, where the OES is exceeded, the employer identifies the reasons for the excess and takes appropriate action to remedy the situation as soon as is reasonably practicable. OESs and MELs are set on the recommendation of the Advisory Committee on Toxic Substances (ACTS) following assessment by the Working Group on the Assessment of Toxic Chemicals (WATCH) of the toxicological, epidemiological, and other data. In 1995, about 40 substances had MELs, about 450 substances had OESs, over 100 substances were listed as carcinogens (some also have MELs), and an additional 72 substances were in the ACTS/WATCH review process. EH40/95 clearly states that OESs are for use only in the workplace and cannot be extrapolated to evaluate and control nonoccupational exposures.

The EC also requires risk assessments for chemicals as part of Council Directive 67/548, Assessment of Notified Substances, which deals with new substances, and Council Directive 793/93, the Evaluation and Control of Existing Substances. Directive 67/548 requires that member states ensure that substances cannot be placed on the market unless certain notification, labeling, and packaging requirements are met. Member states must perform a risk assessment when premarket notification submissions are received. Assessments must be conducted in accordance with the principles set forth in Directive 93/67, which lays down the principles for assessment of risks to man and the environment for substances notified under Directive 67/548. Directive 93/67 specifies the four-step paradigm first articulated by the U.S. National Research Council that is discussed at length in Chapter 2. In addition, the assessment must indicate one or more of the following conclusions:

- the substance is of no immediate concern and need not be reconsidered unless further information is required to be submitted when threshold quantities are met;
- the substance is of concern, but requests for further data will be deferred until such thresholds are met;
- the substance is of concern and further information is needed immediately; or
- the substance is of concern and the competent authorities must make recommendations immediately to reduce the risk.

Annex I to the Directive identifies the toxic effects required to be considered for human health risks (i.e., acute toxicity, irritation, corrosivity, sensitization, repeated dose toxicity, mutagenicity, carcinogenicity, and toxicity for reproduction). The human populations for consideration are workers, consumers, and persons exposed indirectly by way of the environment. Annex II covers risk assessment for human health, including explosivity, flammability, and oxidizing potential. Annex III covers risk assessment for the environment.

Directive 67/548 was implemented in the U.K. by the Notification of New Substances Regulation 1993, which requires persons wishing to place a new substance on the market in a total quantity of one metric ton or more per year to notify the competent authority and provide a dossier supplying "the information, necessary for evaluating the foreseeable risk, whether immediate or delayed, which the substance may create for human health and the environment." Based on the information, the competent authority performs a risk assessment of the new substance. EC published technical guidance notes in supplement Directive 93/67 that was described above. The technical guidance provides detailed guidance on the risk assessment methods, including strategies for toxicity testing, or the use of structure-activity relationships (SAR), or quantitative structure-activity relationships (QSAR). Dose-response assessment is performed according to animal test methods with a recommendation to determine the "no-observed adverse effect level" (NOAEL) rather than a "no-effect" level. For exposure assessments, the guidance provides specific types of exposures and recommends the assessment method. Finally, risk characterization is performed to determine the likelihood that an adverse effect will be caused under reasonably foreseeable conditions of use in the workplace or by consumers.

Council Directive 793/93 applies to the evaluation and control of existing substances. The purpose of this regulation is to provide information to determine whether additional controls should be imposed. The regulation does this by creating a systematic framework for evaluation of existing substances that includes the following steps:

- initial data submissions,
- prioritization of chemicals for full risk assessment,
- performance of full assessments by individual Member States, and
- consideration of further restrictions on the basis of the risk assessment findings.

The regulation also creates an extensive computerized database to facilitate the determination by the Commission and the member states that companies are fulfilling their obligations. Extensive data are required for highest volume substances (e.g., over 1,000 metric tons). The same type of information is required as described above for new substances and special attention is given to substances that are carcinogenic, mutagenic, or toxic to reproduction.

The Commission establishes priorities for assessment. The first priority list was issued on May 25, 1994, and contained 40 chemicals; the second priority list was issued on September 27, 1995, and contained 36 chemicals. Regulators indicate that they expect to identify up to about 50 chemicals each year as priorities for risk assessment. The substances are selected on the basis of factors identified in Article 8 of Regulation 793/93. These include effects of the substance on man or the environment, lack of data on such effects, work carried out by other international bodies, other community regimes relating to dangerous substances, or chronic effects.

Once a substance is added to the priority list, companies which submitted the initial data on it have six months to submit essentially the same data that are required in support of a full notification of a new substance. The data must be submitted to the member state authorities to which the substance has been assigned. The competent authorities must conduct assessments in accordance with Regulation 1488/94 and submit a final report to the Commission. Regulation 1488/94 is similar to Directive 93/67 discussed above. Again, a technical guidance document is available that recommends risk assessment methods similar to those in the guidance for new substances.

Directive 793/93 was implemented in the U.K. by the Notification of Existing Substances (Enforcement) Regulations of 1994. The HSE produced a guide, *How to Report Data on Existing Chemical Substances*, to advise U.K.-based chemical businesses. Also, a government and industry "unified view" on risk assessment was produced by the U.K. Government/Industry Working Group. The guidance recommends that the estimated human dose (EHD) be calculated and compared with a NOAEL to determine the substance's "critical effect," or the observed adverse effect of most concern. The EHD is determined from ecotoxicity and toxicity data, physico-chemical properties, production volumes, use and disposal, environmental fate, and predicted environmental concentration. The process again is similar to that used for new substances. In addition, the Working Group produced a guidance document, *Risk-Benefit Analysis of Existing Substances*, which addresses the requirement that, upon completion of the risk assessment, strategies be suggested for reducing risks. Risk assessment is also considered, although not rigorously implemented, in the U.K. in considering food additives, food packaging, veterinary product residues in food, pesticides, foodstuff, medicinal products, and land planning. In these cases, quantitative risk assessments are recommended but the methods are open to interpretation. Under Directive 85/337, developers of certain projects likely to have significant impacts on the environment are required to submit to the member state's competent authority an environmental statement regarding the project. However, these are all aimed at outdoor, not indoor, environmental concerns. Control of emissions of hazardous substances into the environment consists of emission limits and the submission of environmental impact assessments using the limits as benchmarks.

Environmental impact assessments identify the risk of emissions into the environment and are performed under the U.K. Integrated Pollution Control (IPC) system. Created under the Environmental Protection Act of 1990, the IPC requires those who wish to engage in prescribed processes and to emit prescribed substances to submit an application to Her Majesty's Inspectorate of Pollution (HMIP). The application process must include details of the process and release of emissions and an assessment of the environmental consequences of the process and emissions. Applicants must show that, after the use of the best available techniques not entailing excessive cost (BATNEEC) to prevent, reduce, and/or render harmless releases of harmful substances, they do not exceed limits on emissions. In addition, after a full assessment of the impact on the environment (i.e., the risks), the applicant implements the best practicable environmental option (BPEO), affording the best overall protection to the environment. Limits are set at the EC and U.K. government levels and apply to a large number of air pollutants.

While there has been little use of risk assessment to date, HMIP is producing a lengthy technical guidance on environmental, economic, and BPEO assessment principles for integrated pollution control. Draft guidance recommends that the potential harm of an emission be calculated by comparing the predicted environmental concentration (PEC) to the Environmental Quality Standard (EQS). If the PEC is 80% or more of the EQS, or the process contribution (PC) to the release is 2% or more of the EQS, the HMIP is signaled a priority for control in connection with the application. The direct environmental effect of a substance released to a particular environmental medium then is assessed as the ratio of the PC to the environmental assessment level (EAL). This quantity denotes the environmental quotient (EQ) for that substance in that medium.

Other sources of relevant risk assessment information in Europe include the following:

- standard setting organizations, such as the British Standards Institution (BSI); and
- EC research programs such as an effort by the Joint Research Centre to provide technical and scientific support to EC policies, a research program in the field of environment and climate, a research program in the field of biomedicine and health, research and development in the field of information technologies, and research and development in the field of industrial and materials technologies.

Both the EC and the U.K. are introducing systems to assess the likely costs of legislative and regulatory actions. Particularly, the Environment Act of 1995 in the U.K. requires the performance of a cost-benefit analysis in connection with regulatory actions or inactions. The newly established Environment Agency (which combines the functions of HMIP, the National Rivers Authority, the Waste Regulation Authority, and some functions of HSE) is required to take account of costs and benefits in its actions. The Act does not require specific methods for cost-benefit analysis or pollution assessment but provides that the appropriate Minister give directions.

VI. THE RISK ASSESSMENT PROCESS

Environmental risk assessment initially focused on carcinogenic risks. This occurred largely because of the intense public concern with cancer and the fact that this concern translated into significant nationwide investigation and research into the causes and probabilities of contracting cancer. Additionally, many forms of cancer are believed to be associated with no lower threshold of effect, meaning that there is a finite probability of cancer resulting from any exposure, no matter how low.[9] Thus, there exists a range of cancer risks associated with any range of exposures and a risk assessment can express the probability of occurrence for a specified exposure.

[9] Very low exposures to a carcinogen may not result in cancer in the person's lifetime because the substance may have a low potency or a long latency period (i.e., a time delay) between initial exposure and expression of the disease.

Of course, noncancer health effects also are important. These often are expressed in terms of a safe threshold of exposure whose pass-fail test can obviate the need for a quantitative risk assessment. However, factors such as genetic variability can make this threshold a range rather than a single value and a risk assessment can again express a probability of effect at a specified exposure.

Regardless of the health effect of interest, the process of assessing the human risks of outdoor or indoor exposure to an air pollutant generally is the same. This process was first defined by a Committee of the National Research Council in 1983 in a report requested by Congress entitled *Risk Assessment in the Federal Government: Managing the Process* (NRC 1983). This report will be discussed in more detail in the next chapter, but in general the NRC recommends that an environmental risk assessment consist of one or more of the following four steps:

1. Hazard Identification — Identification of the potential health hazard associated with exposure to humans
2. Dose–Response Assessment — Assessment of the health hazard as a function of varied exposures or doses
3. Exposure Assessment — Assessment of the likely and worst-case exposures to the health hazard
4. Risk Characterization — Quantitative estimation of the human health risks associated with exposures

Some aspects of these four steps can differ depending upon whether the exposures are indoor or outdoor. For example, hazard identification and dose–response assessment are treated similarly for indoor and outdoor pollutants, but certain pollutants are more prevalent or in much greater concentration indoors (e.g., biologicals, radon, and environmental tobacco smoke). In addition, the transport and fate of the pollutant typically is of greater concern outdoors because there are fewer factors indoors influencing fate and transport. Finally, indoor exposures are more easily determined through monitoring because they tend to be more stable and constant; outdoor exposure estimates often rely on mathematical models because of the great cost inherent in attempting to monitor outdoor concentrations over a large geographical area.

One important function of an environmental risk assessment is to identify and attempt to address the uncertainties and natural variations that can occur. Although cancer has been studied for many years, only in a limited number of instances do scientists and doctors know precisely how and why a particular cancer occurs. This results both from the substantial variability in the genetic makeup of humans as well as the large number of different types of cancer and their causes. In addition, exposure to suspected carcinogens must be tested on animals rather than humans. Because of the relatively short life span of most test animals, such tests are typically conducted at concentrations much higher than those to which humans are normally exposed. This introduces uncertainties in extrapolating from animals to humans and also from high to low doses. Some environmental cancers have been observed in humans as a result of workplace exposures, but these generally occurred as a result of exposures much higher than those experienced by the average human and, thus, face the same high- to low-dose questions.

VII. CURRENT INDOOR AIR RISK ASSESSMENT ACTIVITIES

Indoor air quality is receiving increasing attention as researchers realize what should have been intuitive—that people in the U.S. on average spend approximately 90% of their time indoors and only a small fraction of their time outdoors, where air pollution regulatory and risk assessment attention have focused for so long. This knowledge is only amplified as reports and the research conducted in the 1980s and beyond clearly show that humans are significantly exposed to indoor air pollutants and building-related illnesses are increasing.

Assessments have been conducted by the EPA and others of risks associated with indoor air exposures. Many of these focused on environmental tobacco smoke (ETS) and radon, but a number of other potential indoor air pollutants have received attention, including volatile organic compounds, formaldehyde, polycyclic aromatic hydrocarbons, pesticides, and asbestos. The EPA described its activities on indoor air quality, including risk assessment activities, in the 1989 Report to Congress (EPA 1989). In that report, the EPA assessed the health and economic impacts of indoor air pollution and described methods and strategies for controlling indoor air pollution. Importantly, the EPA summarized the knowledge at the time on the major indoor air pollutants, pollutant sources, and potential health effects associated with exposure. This information is presented in Table 1.1 and serves to focus attention on the pollutants, sources, and activities that are more likely to be associated with indoor air risks.

In 1991, the Deputy Administrator of the EPA testified before a House Subcommittee and provided an update on the EPA's "substantial progress in developing an effective response to indoor air pollution" (EPA 1991). He began by emphasizing that the EPA believed that new indoor air legislation was premature at that time, unnecessary from a standpoint of both authority and resources, and potentially disruptive to the momentum already built up. However, he admitted that studies up to that time led to concern that "people are being exposed to levels of pollutants that may be significant in terms of long-term risks of cancer and other chronic effects." He described for Congress the EPA's initial strategy for dealing with indoor air pollution of focusing both on pollutant-by-pollutant analyses, which are necessary for some known and suspected health risks, and on a clustering approach to look more expeditiously at a broader combination of indoor pollutants and sources. This strategy was to be augmented by minimizing human exposure to biological and chemical contaminants indoors, reducing exposure through source control wherever feasible, providing adequate ventilation in all occupied space, and using air cleaning technologies where appropriate. The EPA's current indoor air strategy focuses on building design, construction, operation, and maintenance rather than specific health risks.

Table 1.1   Indoor Air Pollutants, Sources, and Health Effects

Indoor Air Pollutants	Sources	Health Effects
Environmental tobacco smoke (ETS)	Tobacco smoking	Cancer Irritation to mucous membranes Chronic and acute pulmonary effects Cardiovascular effects
Biological contaminants (Viruses, bacteria, molds, insects and arachnid excreta, pollen, animal and human dander)	Outdoors, humans, animals (moist building areas are amplifiers for some)	Infectious diseases Allergic reactions Toxic effects
Volatile organic compounds (VOCs)	Paints, stains, adhesives, dyes, solvents, caulks, cleaners, pesticides, building materials, office equipment	Irritation Neurotoxic effects Hepatotoxic effects Cancer
Formaldehyde (also a VOC)	ETS, foam insulation, particle board, plywood, furnishings, upholstery	Irritation Allergy Cancer
Polycyclic aromatic hydrocarbons (PAHs)	ETS, kerosene heaters, wood stoves	Cancer Irritation Cardiovascular effects Animal data show decreased immune function, atherosclerosis etiology
Pesticides	Pesticide application indoors and outdoors	Neurotoxicity Hepatoxicity Reproductive effects
Asbestos	Asbestos cement, insulation, other building materials	Asbestosis Cancer
Carbon monoxide (CO)	Combustion appliances, ETS, infiltrated exhaust	Increased frequency and severity of angina Decreased work capacity in healthy adults Headaches, decreased alertness, flulike symptoms in healthy adults Exacerbation of cardiopulmonary dysfunction in compromised patients Asphyxiation
Nitrogen dioxide (NO_2)	Combustion appliances, ETS	Decreased pulmonary function in asthmatics Increased susceptibility to infection in animals Effect on pulmonary function in children, perhaps adults Synergistic effects with other pollutants in animals Decreased immune capability, changes in anatomy and function of lung in animals

(continues)

Table 1.1   (continued)

Indoor Air Pollutants	Sources	Health Effects
Sulfur dioxide (SO$_2$)	Combustion of fuels containing sulfur	Decreased lung function in asthmatics (in synergism with particles increased [doubled] airway resistance) Decreased lung function in animals
Particulate matter	Combustion appliances, ETS	Cancer (soot, PAH adsorbed to particles) Irritation of respiratory tissues and eyes Decreased lung function alone and in conjunction with SO$_2$
Radon	Soil, well water, some building materials	Cancer
Dust sprays, cooking aerosols	Personal activity	Unknown (can range from irritation to cancer)

Source: EPA 1989.

Risk assessment was not being ignored but the indoor air strategy recognized that there was significant uncertainty at that time both in assessing risks and in understanding the meaning of risk assessments. Notwithstanding, because of the closed environment nature of indoor air exposures, a building-oriented strategy appeared to be prudent regardless of the actual estimated risks. The 1989 EPA Report to Congress on Indoor Air Quality focused its risk assessment discussion on carcinogens. It outlined the risk assessment process recommended by the National Research Council in 1983, discussed uncertainties and variability, and summarized the knowledge at that time of cancer risk estimates for typical indoor air pollutants. Since that time, however, two additional important documents dealing with risk assessment have been published.

In 1994, at the request of Congress, the National Research Council published a new report (NRC 1994) entitled *Science and Judgment in Risk Assessment*. This report described the risk assessment process as it was currently conducted and made numerous recommendations for improving the process. Even more recently, the Risk Assessment and Risk Management Commission was formed, as a result of section 303 of the 1990 Clean Air Act Amendments, for the purpose of advising Congress on the policy implications and appropriate uses of risk assessment. Their report in early 1997 proposed a new risk-management framework for decision making, described the uses and limitations of risk assessment and risk management, and provided limited specific recommendations to several federal departments and agencies. While not specifically addressing any specific environmental pollutants or exposure scenarios, these documents will influence the conduct of risk assessment throughout the EPA in the future.

VIII. COMPARISON OF INDOOR AIR RISKS AND OTHER ENVIRONMENTAL RISKS

Few studies have attempted to compare directly and comprehensively environmental risks from indoor air and other sources of exposure. The most notable was the EPA's Unfinished Business Report (EPA 1987) which concluded, although admittedly using imperfect data, that indoor air represents one of the most important environmental problems based on estimated population risks. The EPA's 1989 report to Congress on Indoor Air Quality concluded that "indoor air pollution represents one of the most important environmental problems based on population risks" (EPA 1989). While these risks are generally less than many occupational health and safety risks in mining and industrial environments, and some ecological and welfare losses for some environmental problems may dominate in public importance, the report to Congress also stated that "population health risks posed by exposure to indoor air pollutants appear to be significantly greater than the health risks posed by some of the environmental problems that receive the most public concern and governmental funding."

IX. LEGISLATIVE AND REGULATORY INITIATIVES ADDRESSING INDOOR AIR AND RISK ASSESSMENT

A. U.S. Federal

1. National Environmental Policy Act (42 USC 4321)

The National Environmental Policy Act (NEPA) was the first major environmental legislation enacted and requires for any proposed legislation or proposed major federal action, including state and local action using federal funding and actions where federal approval is required, that the effects on the quality of the human environment be assessed. All environmental consequences, resource uses, and reasonable alternatives must be identified and considered as part of a public decision-making process. While not specific to indoor air, the legislation is sufficiently broad to require consideration of indoor air quality; however, risk assessment is not addressed specifically as a tool in the analysis nor is it typically used.

2. U.S. Environmental Protection Agency

The EPA currently maintains an indoor air quality program designed to: (1) characterize indoor air problems, (2) identify, assess, and implement strategies to mitigate indoor air hazards, and (3) disseminate information about indoor air quality control. These activities are coordinated by the Office of Radiation and Indoor Air in the Office of Air and Radiation. This group, in conjunction with the EPA's Office of Research and Development, develops reports and informational materials to assist other regulators and the public in evaluating and dealing with indoor air quality. One of the most widely distributed reports is *Building Air Quality — A Guide for Building Owners and Facility Managers*, published in 1991 (EPA 1991). The EPA also publishes voluntary guidelines, disseminates public information, and conducts research related to indoor air quality. The EPA activities under specific legislation are summarized below.

a. Clean Air Act (42 USC 7401)

As described above, the Clean Air Act (CAA) addresses two major types of air pollutants that can be of concern both outdoors or indoors, and it serves as the vehicle for risk-based regulations. However, the CAA was written and generally is interpreted as applying to outdoor air although reduction in concentrations of outdoor air pollutants will result in a proportionate reduction in infiltration of those pollutants indoors.

b. Toxic Substances Control Act (15 USC 2601)

The Congressional intent of the Toxic Substances Control Act (TSCA) was to require manufacturers and processors of chemical substances and mixtures to control those substances in such a way as to prevent unreasonable risks of injury to health and the environment. This act provides the EPA with the authority to restrict the manufacture, distribution, and use of toxic chemicals. Presumably, it could be applied to indoor air pollutants, although to date the Agency has not done so.

c. Asbestos Regulations

The Asbestos School Hazard Abatement Act (Title V of Public Law 98-377) was enacted in 1984 requiring the EPA to establish programs to assist state and local agencies in identifying and abating asbestos in schools. The Asbestos Hazard Emergency Response Act (AHERA) (15 USC 2641) was passed as Title II of the Toxic Substances Control Act of 1986. AHERA deals with the specific issue of asbestos in school buildings and it establishes the programs for evaluation, mitigation, and accreditation; however, it does not address risk assessment. Finally, the Asbestos Information Act of 1988 (Public Law 100-577) requires manufacturers and processors of asbestos or asbestos-containing materials to provide certain information to the EPA on their products. These laws led to a variety of federal requirements dealing with asbestos in public buildings, the workplace, and the ambient environment.

d. SARA Title IV (42 USC 7401)

As part of the 1986 Superfund Amendments and Reauthorization Act (SARA), Congress included Title IV — Radon Gas and Indoor Air Quality Research. Congress expressed concerns with the serious health threats posed by radon in structures in certain areas of the country and established Title IV to develop an adequate information base concerning radon and indoor air pollutants. It directed the EPA to: establish a research program to gather information on all aspects of indoor air quality; coordinate with federal, state, local, and private research and development; and assess appropriate governmental mitigation measures. Specific research priorities were identified, including: research and development concerning the identification, characterization, and monitoring of sources and levels of indoor air pollution; research relating to the effects of indoor air pollution and radon on human health; research and development relating to control technologies or other mitigation measures to prevent and abate indoor air pollution; demonstration of methods for reducing and eliminating indoor air pollution and radon; research, in conjunction with the Department of Housing and Urban Development, to develop methods for assessing the potential for radon contamination in new construction and design measures to avoid indoor air pollution; and dissemination of information to assure the public availability of the findings. Again, there was no direct requirement for risk assessment.

e. Safe Drinking Water Act (42 USC 300f)

The Safe Drinking Water Act (SDWA) authorizes the EPA to conduct research and to establish standards to protect the safety of the U.S. drinking water systems. While not specifically directed at the indoor air, a number of substances in drinking water are regulated that can affect indoor air quality. Some of the more widely studied are radon and halomethanes that can volatilize from hot water, particularly in showers. Other concerns are biological contaminants that can enter the indoor environment either directly in drinking water or indirectly through heating and cooling systems. Establishment of drinking water criteria under the SDWA can involve an evaluation of the risks associated with exposure.

f. Federal Insecticide, Fungicide, and Rodenticide Act (7 USC 136)

Pesticide standards are promulgated under the Federal Insecticide, Fungicide, and Rodenticide Act (FIFRA), several of which are aimed specifically at indoor air exposures. For example, bans or restrictions in place on chlordane, heptachlor, lindane, pentachlorophenol and creosote are based in part on the potential for indoor exposures. Standards under FIFRA can involve use of risk assessment.

3. U.S. Occupational Safety and Health Administration

The Occupational Safety and Health Administration (OSHA) is responsible for regulating health and safety in the workplace. Importantly, the OSHA administers the Occupational Safety and Health Act (29 USC 651), enacted in 1970 "to assure as far as possible every working man and woman in the Nation safe and healthful working conditions and to preserve our human resources. . . ." The OSHA's workplace air standards (29 CFR 1910.1000) represent the largest source of indoor air quality criteria. Although the subject of ongoing controversy, these standards are generally being complied with and reduce workplace exposures to nearly 400 individual chemicals and substances. In April 1994, the OSHA also proposed new workplace standards on indoor air quality prompted largely by petitions relating to environmental tobacco smoke. The proposal was based on the OSHA's stated determination that employees in indoor working environments face a significant risk of material impairment to their health due to poor indoor air quality. The proposal was far reaching and attracted over 100,000 comments and over 400 witnesses in public hearings. At the time of this writing, the OSHA continued to review the comments and testimony and no date was set for further action.

4. U.S. Department of Energy

The U.S. Department of Energy (DOE) has broad authority to manage U.S. energy-related programs. A stated purpose of the DOE Organization Act of 1977 was to "assure incorporation of national environmental protection goals in the formulation and implementation of energy programs, and to advance the goals of restoring, protecting and enhancing environmental quality, assuring public health and safety." Because of the close relationship between energy management and air quality both indoors and outdoors, many studies have been and are being conducted using DOE funding that directly relate to indoor air quality; however, risk assessment typically is not prescribed.

5. U.S. Department of Health and Human Services

The U.S. Department of Health and Human Services (DHHS) has the broad responsibility of protecting the public health and preventing disease; as such, many of their actions and research relate to indoor air quality. In particular, the Public Health Service Act (PHSA) requires the identification of pollution and environmental conditions "responsible for human disease and adverse effects in humans," and it authorizes studies, in cooperation with other federal agencies and entities, of the health costs of pollution and other environmental activities resulting from activities "including human activity in any place in the indoor or outdoor environment, including places of employment and residences." While risk of pollution and environmental conditions responsible for disease and adverse health effects are an inherent consideration in these studies, risk assessment is not specifically required.

6. U.S. Consumer Product Safety Commission

The Consumer Product Safety Commission (CPSC) is an independent regulatory agency established to protect consumers from unreasonable risks of injury. Acting under the Consumer Product Safety Act (CPSA) and the Federal Hazardous Substances Act (FHSA), the CPSC can ban consumer products from the market, but their authority aims principally at manufacturers of products. Examples are the ban on carbon tetrachloride and asbestos in consumer products and the ban on vinyl chloride use in aerosol products. Risk assessment is not a primary tool in the decision process. The CPSC also publishes voluntary guidelines, disseminates public information, and funds research directed at indoor air quality issues.

7. U.S. Department of Housing and Urban Development

The Department of Housing and Urban Development (HUD) administers the nation's housing and urban development policies through a variety of financial and technical assistance programs. A stated policy in HUD's enabling legislation is to "provide decent, safe and sanitary housing." While indoor air quality is not a factor in decision making, in promulgating regulations under the National Manufactured Housing Construction and Safety Standards Act of 1974, HUD included formaldehyde emission controls for certain wood products and outdoor air ventilation requirements for manufactured housing using forced heat. However, risk assessment is not specified as a tool in the analysis or control.

B. Others in the U.S. Involved in Indoor Air Quality

1. State and Local Regulatory Agencies

Indoor air quality is regulated in some way by a majority of the states in the U.S. A large number have placed restrictions on smoking in public spaces and many of them incorporate air quality considerations into building codes. Asbestos and radon are also widely regulated; however, few states have promulgated limits on indoor air concentrations or emissions of specific substances, although some states have incorporated the OSHA standards for nonindustry workplaces. Many states also disseminate information and fund research related to indoor air quality.

2. Private Organizations

A wide variety of private organizations contribute to the improvement of indoor air quality knowledge. These include independent standards-setting organizations such as the following:

- American Conference of Governmental Industrial Hygienists (ACGIH)
- American Society of Heating, Refrigeration, and Air Conditioning Engineers (ASHRAE)
- American Society for Testing and Materials (ASTM)
- Underwriters Laboratory (UL)

These and other organizations provide consensus guidance and standards for indoor air quality that are accepted and complied with widely. However, these organizations do not typically use risk assessment as a tool in their recommendations and guidance.

C. International Organizations

1. Countries

A number of countries deal with indoor air quality issues but few use risk assessment directly as a tool in decision-making. For example, Canada controlled urea-formaldehyde foam insulation, required ventilation rate changes in their building code, and issued guidelines on acceptable indoor air concentrations for a limited number of substances. The United Kingdom directly regulates indoor air quality and publishes voluntary guidelines. Specific programs are in place for asbestos, formaldehyde, and combustion products. Scandinavian countries conduct extensive research on indoor air quality problems and several have specific ventilation requirements.

2. World Health Organization

Since before 1980, the World Health Organization (WHO) has played a leading role in Europe on indoor air quality issues. Among their activities are research and communication programs for indoor air quality and guidelines for pollutant concentrations in indoor and outdoor atmospheres. WHO also sponsors or cosponsors international conferences addressing indoor air quality and climate issues.

BIBLIOGRAPHY

Air and Waste Management Association. 1991. *Air Toxics Issues in the 1990s. Policies, Strategies and Compliance, Proceedings from an Air and Waste Management Association International Conference*, Air and Waste Management Association: Pittsburgh, PA.

American Conference of Governmental Industrial Hygienists. 1989. *Guidelines for the Assessment of Bioaerosols in the Indoor Environment*, American Conference of Governmental Industrial Hygienists: Cincinnati, OH.

American Industrial Hygiene Association. 1994. Comments on OSHA's proposed indoor air quality standard, *59 FR 15968–16039*.

Brown, K.S. 1996. Sick days at work, *Environmental Health Perspectives* 104(10):1–8 (on-line).

Brown, R.C., Hoskins, J.A. 1992. Contamination of indoor air with mineral fibres, *Indoor Environment* 1:61–68.

Bureau of National Affairs. 1989. *Asbestos Abatement Resource Guide*, Second Edition, Bureau of National Affairs: Washington, DC.

Burge, H.A. 1990. Risks associated with indoor infectious aerosols, *Toxicology and Industrial Health* 6(2):263–274.

Burge, H.A. 1995. *Bioaerosols, Indoor Air Research Series*, Center for Indoor Air Research, Boca Raton, FL: Lewis Publishers.

Burr, M.L. 1995. Pollution: Does it cause asthma?, *Journal of the British Paediatric Association* 72(5):377–379.

Carswell, F. 1995. Epidemiology of the effects of air pollutants on allergic disease in children, *Clinical and Experimental Allergy* 25(3):52–55.

Chemical Manufacturers Association. 1988. *Chemicals in the Community: Methods to Evaluate Airborne Chemical Levels,* Chemical Manufacturers Association: Washington, DC.

Cohen, L. 1995. Indoor air quality a major public-health issue, workshop told, *Canadian Medical Association Journal* 153(1):92–93.

Covington & Burling. 1995. Memorandum Re: Methods of Risk Assessment in the European Community and the United Kingdom, November 29, 1995.

Department of Health and Human Services (DHHS). 1985. *Risk Assessment and Risk Management of Toxic Substances,* A Report to the Secretary, Department of Health and Human Services from the DHHS Committee to Coordinate Environmental and Related Programs (CCERP), April, 1985.

Environ. 1986. Elements of Toxicology and Chemical Risk Assessment, A Handbook for Nonscientists, Attorneys and Decision Makers, Environ Corporation: Washington, DC.

Environmental Protection Agency (EPA). 1979. National emission standards for identifying, assessing and regulating airborne substances posing a risk of cancer, *44 FR 58642.*

Environmental Protection Agency (EPA). 1981. *Policy and Procedures for Identifying, Assessing and Regulating Airborne Substances Posing a Risk of Cancer,* Draft, 6/22/81, Pollutant Assessment Branch, Strategies and Air Standards Division, Office of Air Quality Planning and Standards, U.S. Environmental Protection Agency.

Environmental Protection Agency (EPA). 1984. *Risk Assessment and Management: Framework for Decision Making,* Report No. EPA 600/9-85-002, U.S. Environmental Protection Agency, Washington, DC.

Environmental Protection Agency (EPA). 1985a. *Bibliographic Series,* Report No. EPA/IMSD-85-002, Information Services and Library, U.S. Environmental Protection Agency, Washington, DC.

Environmental Protection Agency (EPA). 1985b. *The Air Toxics Problem in the United States: An Analysis of Cancer Risks for Selected Pollutants,* Report No. EPA 40/1-85-001, Office of Air and Radiation, Office of Policy, Planning and Evaluation, U.S. Environmental Protection Agency, Washington, DC.

Environmental Protection Agency (EPA). 1987a. *Unfinished Business: A Comparative Assessment of Environmental Problems. Appendix II Non-Cancer Risk Work Group,* Office of Policy Analysis, Office of Policy, Planning and Evaluation, U.S. Environmental Protection Agency, Washington, DC.

Environmental Protection Agency (EPA). 1987b. *Unfinished Business: A Comparative Assessment of Environmental Problems. Appendix III Ecological Risk Work Group,* Office of Policy Analysis, Office of Policy, Planning and Evaluation, U.S. Environmental Protection Agency, Washington, DC.

Environmental Protection Agency (EPA). 1987c. *Unfinished Business: A Comparative Assessment of Environmental Problems. Appendix IV Welfare Risk Work Group,* Office of Policy Analysis, Office of Policy, Planning and Evaluation, U.S. Environmental Protection Agency, Washington, DC.

Environmental Protection Agency (EPA). 1987d. *Unfinished Business: A Comparative Assessment of Environmental Problems,* Office of Policy, Planning and Evaluation, U.S. Environmental Protection Agency, February, 1987.

Environmental Protection Agency (EPA). 1989a. *Report to Congress on Indoor Air Quality,* Report No. EPA/400/1-89/001A, Office of Air and Radiation and Office of Research and Development, U.S. Environmental Protection Agency, August.

Environmental Protection Agency (EPA). 1989b. *Risk Assessment Guidance for Superfund, Volume I Human Health Evaluation Manual (Part A), Interim Final,* Report No. EPA/540/1-89/002, Office of Emergency and Remedial Response, U.S. Environmental Protection Agency, Washington, DC.

Environmental Protection Agency (EPA). 1990a. *Building Air Quality: A Guide for Building Owners and Facility Managers*, Office of Air and Radiation, U.S. Environmental Protection Agency, Washington, DC.

Environmental Protection Agency (EPA). 1990b. *Managing Asbestos in Place, A Building Owner's Guide to Operations and Maintenance Programs for Asbestos-Containing Materials*, Report No. 20T-2003, Pesticides and Toxic Substances, U.S. Environmental Protection Agency.

Environmental Protection Agency (EPA). 1991a. *Building Air Quality: A Guide for Building Owners and Facility Managers*, U.S. Environmental Protection Agency published in cooperation with the U.S. Department of Health and Human Services, ISDN 0-16-035919-8, December.

Environmental Protection Agency (EPA). 1991b. *Risk Assessment for Toxic Air Pollutants: A Citizen's Guide*, Report No. 450/3-90-024, Office of Air and Radiation, U.S. Environmental Protection Agency, Washington, DC.

Environmental Protection Agency (EPA). 1991c. *Risk Assessment Guidance for Superfund: Volume I-Human Health Evaluation Manual (Part B, Development of Risk-Based Preliminary Remediation Goals)*. Publication No. 9285.7-01B, Office of Emergency and Remedial Response, U.S. Environmental Protection Agency, Washington, DC.

Environmental Protection Agency (EPA). 1991d. Testimony of Deputy Administrator F. Henry Habicht, Jr. before the Subcommittee on Health and the Environment, U.S. House of Representatives, April 10.

Environmental Protection Agency (EPA). 1992. *Safeguarding the Future: Credible Science, Credible Decisions*, Report No. EPA/600/9-91/050, U.S. Environmental Protection Agency.

Environmental Protection Agency (EPA). 1996. Proposed Guidelines for Carcinogen Risk Assessment, Report No. EPA/600/P-92/003C, Office of Research and Development, U.S. Environmental Protection Agency.

Gammage, R.B. 1994. *Resolving the Ambiguities: An Industrial Hygiene Indoor Air Quality (IAQ) Symposium*, American Industrial Hygiene Association.

General Accounting Office. 1983. *Report by the Comptroller General of the United States. Delays in EPA's Regulation of Hazardous Air Pollutants*, U.S. General Accounting Office, Washington, DC.

Health and Safety Commission. 1995. *General COSHH ACOP and Carcinogen ACOP and Biological Agents ACOP, Control of Substances Hazardous to Health Regulations 1994, Approved Codes of Practice*.

Health and Safety Executive. 1995. *EH40/95 Occupational Exposure Limits 1995*, HSE Books: Sudbury, England.

Jayjock, M.A., Hawkins, N.C. 1993. A proposal for improving the role of exposure modeling in risk assessment, *American Industrial Hygiene Association Journal* 54(12):733–741.

Lowry, S. 1989. Indoor air quality, *British Medical Journal* 299:1388–1390.

MEDACCESS. 1996. Indoor air pollution: An introduction for health professionals (on-line).

National Research Council (NRC). 1983. *Risk Assessment in the Federal Government: Managing the Process*, Prepared by the Committee on the Institutional Means for Assessment of Risks to Public Health, Commission on Life Sciences, National Academy Press: Washington, DC.

National Research Council (NRC). 1994. *Science and Judgment in Risk Assessment*, Prepared by the Committee on Risk Assessment of Hazardous Air Pollutants, Commission on Life Sciences, National Academy Press: Washington, DC.

Occupational Safety and Health Administration (OSHA). 1996. Preamble and Proposed Regulations for Workplace Indoor Air, *29 CFR 1910, 1915, 1926, 1928*, Occupational Safety and Health Administration.

Özkaynak, H., Spengler, J.D. 1990. Introduction to the risk assessment workshop on indoor air quality, *Risk Analysis* 10(1):15–17.

Patrick, D.R. *Risk Assessment and the New Clean Air Act: The Good News and the Bad News,* ICF Kaiser Engineers.

Ruckelshaus, W.D. 1983. *Science, Risk, and Public Policy,* National Academy of Sciences, June 22, 1983.

Tibbetts, J. 1996. Green houses, *Environmental Health Perspective* 104(10):1–8 (on-line).

Volkmer, R.E., Ruffin, R.E., et al. 1995a. The prevalence of respiratory symptoms in South Australian preschool children. I. Geographic location, *Journal of Paediatric Child Health*, 31:112–115.

Volkmer, R.E., Ruffin, R.E., et al. 1995b. The prevalence of respiratory symptoms in South Australian preschool children. II. Factors associated with indoor air quality, *Journal of Paediatric Child Health*, 31:116–120.

Walter, M. 1995. OSHA's IAQ ruling in doubt as industry vents concerns, *Environmental Solutions*, August, 1995:48–50.

World Health Organization. 1996. *The Tobacco Epidemic: A Global Public Health Emergency, World No-Tobacco Day 1996 Special Issue*, Programme on Substance Abuse, World Health Organization: Geneva, Switzerland.

CHAPTER **2**

The Elements of Human Health Risk Assessment

Elizabeth L. Anderson and David R. Patrick

CONTENTS

I. INTRODUCTION

Risk assessment is the systematic evaluation of the factors that might result in an adverse human health effect resulting from exposures to contaminants and often the attempted quantification of those factors and effects. The expression of a human health risk is dependent upon two principal components—toxicity and exposure. In other words, a human must be exposed to a substance that can cause an adverse health effect for there to be a risk. The purpose of the human health risk assessment is to evaluate these components, to estimate the likelihood that the adverse health effect might occur, and to determine the magnitude of the associated impact.

This chapter broadly describes the science of risk assessment as it exists today and how it began, and then briefly defines the key steps and data needs for a human health risk assessment. The major steps in the risk assessment are treated in more detail in the following chapters. The purpose of this chapter is to acquaint the reader with the terminologies in this book before delving more deeply into the subject.

II. THE BEGINNINGS OF HUMAN HEALTH RISK ASSESSMENT

Public concern over environmental pollution grew rapidly in the 1960s and 1970s. As public health officials considered potential environmental problems and their possible solutions, it became clear that many, if not most, environmental pollutants could not be regulated or controlled to levels at which there was no risk to the population or the environment. Some substances and their sources might be eliminated, but a healthy and growing economy meant that many suspect contaminants would continue to be produced, used, or released into the environment, and that at least some humans and some part of the environment would inevitably be exposed. Scientists and public health officials thus began searching for methods that would enable rational and prudent decision-making about which of these potential hazards and exposures should be reduced and by how much. In other words, methods were needed that allowed a public health official to determine when the environmental risks associated with an activity were of such magnitude that the adverse effects outweighed the benefits of the activity to society.

As described in Chapter 1, these procedures were in development for some years and there were attempts to apply them in some regulatory programs. However, it was not until 1983 that a procedure with broad scientific consensus was presented in the U.S. for conducting and using environmental risk assessment in the public health decision-making process. That procedure, published at the request of Congress by a Committee of the National Research Council (NRC), established the basic framework for federal regulatory decision making using risk assessment (NRC 1983). Two important concepts came out of this publication:

1. The Committee codified the process by which environmental risk assessments should be conducted. They stated that a risk assessment should contain some or all of the four steps. The four steps, usually referred to as the *risk assessment paradigm*, are hazard identification, dose–response assessment, exposure assessment, and risk characterization.
2. The Committee articulated the need for a clear conceptual distinction between risk assessment and risk management. In this distinction, the scientific findings and policy judgments embodied in risk assessments should be explicitly distinguished from the political, economic, and technical considerations that influence the design and choice of regulatory strategies.

This framework is now largely embraced by regulatory agencies across the U.S. and in many other countries and serves to guide current decision making on environmental risks. For example, in 1983 then Administrator Ruckelshaus announced that the EPA accepted the Committee's recommendations and he committed the Agency to using the procedure where appropriate in its decision making and to involving the public more fully in the decision-making process (Ruckelshaus 1983).

The NRC recognized that the paradigm was not a panacea for all risk assessment questions. There are and almost always will be substantial scientific uncertainties in environmental risk assessment. In addition, making policy and public health decisions using uncertain risk estimates is a formidable task. Nonetheless, scientists and regulators do have a tool, however blunt, that helps them organize available information and provides a framework for responsible decision making. As discussed in Chapter 1, useful insights into the rational use of risk assessment and risk management were added recently (NRC 1994; Commission on Risk Assessment and Risk Management 1997). While a complete risk assessment/risk management process for indoor air pollutants is not yet codified, there is a growing prospect for a decision tool in the foreseeable future that will be widely accepted and used.

III. THE RISK ASSESSMENT PARADIGM

The 1983 NRC report provided that a human health risk assessment should contain some or all of the following four steps (Moschandreas [1988] provides useful explanatory material relating to indoor air quality):

Hazard Identification. Hazard identification is the determination whether a particular substance is or is not causally related to a particular health effect. Hazard identification determines whether exposure to a contaminant causes an adverse effect. It does not seek quantitative results but requires review of all relevant data including epidemiology, animal bioassay, physical and chemical structure, and *in vitro* research.

Dose–Response Assessment. Dose–response assessment is the determination of the relation between the magnitude of exposure and the probability of occurrence of the health effect in question. It establishes a quantitative relationship between the dose administered and the response (i.e., health impact) in humans using studies that may involve epidemiology or animal test data. Both contain uncertainties and models generally are required to extrapolate from high experimental doses to low ambient doses and from animals to humans.

Exposure Assessment. Exposure assessment is the determination of the extent of human exposure before and after application of regulatory controls. Exposure assessment quantifies human exposure to a contaminant and estimates the impact of changing conditions. Exposure to an air contaminant is the integration of pollutant concentrations and the times the humans are exposed to those levels. Both measurement and mathematical models are used outdoors to estimate pollutant concentrations as a function of emission rate; in the more stable indoor environment, measurement is often adequate and generally more accurate.

Risk Characterization. Risk characterization is the description of the nature and often the magnitude of the human risk, including attendant uncertainties. Risk characterization is the synthesis of the information from the other three steps. It usually estimates the incidence rate of an adverse health effect associated with the contaminant of concern. Risk characterization is also the communication link for transferring information to the policy makers, who combine it with other economic, social, and political inputs to reach a decision, take action, and communicate the results to the public.

Each of these risk assessment steps is associated with uncertainties. As discussed in Chapter 7, uncertainty in risk assessment includes both scientific uncertainty and variability. In a risk assessment, these can take several forms, including:

- model variations,
- model input variations,
- lack of complete knowledge of the underlying science, and
- natural variation.

The first three of these uncertainties can often be reduced by gathering additional data and conducting research studies; the fourth generally cannot be reduced but often can be reasonably estimated. In each area, assumptions are often required. The way these assumptions are selected is usually determined by the underlying reason for the assessment combined with numerous scientific, policy, and resource considerations. For example, a priority-setting study might use very conservative assumptions to ensure that all possible candidates for priority setting are identified. On the other hand, a regulatory decision, with a potential for substantial health risks or control costs, will generally attempt to use assumptions intended to quantify the risks and exposures as accurately as possible. These areas of uncertainty are discussed more fully in Chapter 7.

IV. THE APPLICATION OF THE RISK ASSESSMENT PARADIGM TO INDOOR AIR

A. Hazard Identification

The hazard identification step is generally the same for outdoor and indoor pollutants; it involves a determination of whether adverse health effects are associated with exposure to a specific substance. This determination can involve: gathering physical and chemical properties of the substance; conducting or evaluating toxicological studies on animals, humans, or other laboratory species; gathering metabolic and physiological data on animals and humans; and investigating likely pathways of exposure. All of the information is then evaluated and the likelihood of an adverse effect from exposure is determined. This is often called the "weight of evidence." Currently, there is no agreed upon means for quantifying weight of evidence. One important part of this step is the determination of how specific exposures (e.g., through inhalation, ingestion, or skin contact) can result in an adverse dose. In other words, the assessor tries to determine how a substance is integrated into the body so that it might result in an adverse effect.

Özkaynak and Spengler (1990) note an important role for hazard identification in the future in accounting for multipollutant impacts by mixture analysis. Nowhere is this more important than in the assessment of indoor air pollution. In the past, hazard identification for environmental regulations focused on individual pollutants rather than total exposure. This resulted both from the media-based legislation (i.e., laws for air, water, and solid and hazardous wastes) and the industrial focus of the environmental legislation. While people are often exposed to individual pollutants at relatively high concentrations, typical outdoor and indoor exposures are to mixtures of chemicals at low concentrations. The EPA published guidelines for dealing with chemical mixtures (EPA 1986d), but information relating to the effects of mixtures on humans comes largely from occupational observations. The Commission on Risk Assessment and Risk Management discussed above recommended increased toxicity testing of complex environmental mixtures of regulatory importance. While they did not specifically mention indoor air pollution, the indoor environment provides the opportunity for more concentrated and continuous exposures to mixtures of pollutants than are generated both indoors and outdoors.

To reduce the uncertainties of animal and laboratory testing, epidemiology studies often are conducted. As described in Lipfert (1994), epidemiology is the study of variations in the incidences of disease or the states of well-being. These studies are generally descriptive or analytic. Descriptive studies generally investigate disease rates in populations in comparison to temporal or spatial distribution of the suspected risk factors; analytic studies generally investigate individuals or groups of individuals in comparison to reference populations. Epidemiologic studies can be powerful tools for identifying hazards; however, they rely on statistical evaluation of enormous quantities of data, some of which can be associated with potentially significant bias and confounding from other risk factors. Normally, epidemiologic studies must be used in combination with other data to establish causality. In addition, an epidemiologic assessment of health effects in a subject population should identify sensitive population groups. Differences in sensitivity can result from age, sex, genetic, nutrition, and life-style differences, as well as the presence of other diseases in the population.

Animal studies are conducted to predict effects in humans and to provide insights into mechanisms of action. These tests are conducted by governmental agencies as well as universities and private organizations. Appropriate experimental design has been established through a scientific consensus process developed over many years. In general, these studies seek to determine the no-observed-adverse-effects-level (NOAEL) and the lowest-observed-adverse-effects-level (LOAEL). Three levels of studies are typically performed: acute, subchronic, and chronic. Acute effects occur from short-term exposures (typically up to a few hours); subchronic effects occur from exposures that are intermediate in nature and nonlethal (typically up to a few months); and chronic effects occur from long-term exposures (typically a substantial portion of the animal's life span). In all cases, the differences in the responses of animals to the exposures typically result in substantial uncertainties in extrapolation of the results to humans. More directed animal studies may also be performed, including carcinogenicity, developmental toxicity, reproductive toxicity, neurotoxicity, and genotoxicity.

Finally, the hazard identification step attempts to determine what the EPA calls the "weight-of-evidence." This is the qualitative assessment of toxicity data and the judgment that exposure to a particular substance is or is not causally related to the expression of an adverse health effect in humans.

B. Dose–Response Assessment

The dose–response assessment step looks beyond the hazard identification and attempts to determine the specific responses that can occur at varying doses. Özkaynak and Spengler (1990) discuss dose–response needs including: biological factors such as body weight, breathing rate, diet and personal habits; intake and dose data on exposure pathways such as inhalation, ingestion of water, milk, food, and soil, and skin contact; and the incorporation of the potential effects of various time-activity patterns of the different segments of the population.

In the past, it was generally assumed that substances have two fundamentally different toxicological mechanisms of action. Some substances were assumed to have an exposure threshold below which there is apparently no adverse effect; other substances were assumed to have no exposure threshold and to result in the potential for an effect at any dose. Most environmental pollutants that were not associated with a potential for carcinogenicity or mutagenicity were categorized as threshold substances. More recently, the report by the Commission on Risk Assessment and Risk Management discusses how recent scientific evidence shows that this distinction has become blurred. For example, an early assumption that all carcinogens are mutagens is inconsistent with current scientific knowledge; similarly, some pollutants normally treated as threshold pollutants (e.g., ozone) may not have a definable threshold for some adverse effects.

The dose–response step involves conducting or evaluating laboratory studies, or conducting or evaluating the effects of controlled exposures on animals (in the laboratory) and humans (in the laboratory or workplace). However, these studies are limited by considerations such as the following:

- animal studies normally must be conducted at relatively high concentrations because of the short life span of the typical animal test subject;
- exposures to humans in the workplace typically are much higher than those experienced by the average person in the ambient environment and they typically are episodic rather than continuous; and
- studies of humans in the laboratory can only be conducted where there is assurance that any adverse effects will not be permanent or debilitating.

In view of these limitations, the risk assessor must often make assumptions, such as the extrapolation of results from animals to man and from high to low doses.

Each of these assumptions can be the subject of considerable scientific debate. The dose–response step works in conjunction with the hazard identification step to attempt to express both the weight of the evidence (i.e., the likelihood that exposure to a substance can result in an adverse effect) and the potency (i.e., the dosage required to produce an adverse effect).

Because of its prevalence, cancer has received the most attention in the attempts to quantify dose–response relationships. However, even after so many years of research, there are still many more questions about the causes of cancer than there are answers. A major reason is that cancer is not one disease, but a large number of related diseases, many of which have unique causes and cures. As understood today, most forms of cancer are believed to begin with an initiation step involving a change in genetic material. To become cancer, however, promotion must occur. In this step, which is dependent upon the rates of repair and cell division, a new transformed cell is produced. Finally, the transformed cells can become malignant tumors. The understanding of the carcinogenic dose–response process must usually rely on animal tests because only a few substances in the environment are positively associated with cancer in humans, principally because of the high background rate of cancer in the population. A variety of methods have been proposed to facilitate the extrapolation of test data to humans. More recently, pharmacokinetic methods are being used to better determine the dose–response relationship by attempting to determine the biologically effective dose of a substance reaching the target organ. These efforts can involve a wide range of bodily functions and processes down to the cellular level. Current research is also finding that some cancers do not occur through these processes and additional dose–response models are being developed to properly assess them.

Noncarcinogenic dose–response assessment can involve many of the same issues as carcinogens. Most noncarcinogens are viewed as having a definable threshold of effect, meaning that exposure from zero up to a finite threshold can be tolerated without adverse effect. The establishment of a dose–response relationship often involves the application of uncertainty factors to human or animal test results. The appropriate factors result from a scientific consensus established over many years. For example, factors may be applied as a result of the extrapolation from humans to sensitive humans, animals to humans, subchronic to chronic, LOAEL to NOAEL, and in the use of incomplete data.

C. Exposure Assessment

The exposure assessment step involves the estimation of the magnitude, duration, and route of exposure to a substance. In the early years of risk assessment, exposure assessments were typically limited to a single substance and evaluation of the most direct exposure pathway (i.e., inhalation for air pollutants, ingestion for water pollutants, and skin contact for soil pollutants). Assessors now recognize that many exposures involve multiple pathways and that humans can be exposed to many pollutants at once. For example, a person living near an industrial source of a pollutant can be exposed by inhaling the air containing source emissions, eating home grown vegetables that have absorbed deposited air pollutants, eating animals or fish contaminated by pollutants released from the source, and drinking ground-water contaminated by chemical leaks from the source. That same person can be exposed indoors to a variety of other chemicals both emitted in the indoor environment and brought in or infiltrated from the outdoors. Energy conservation measures may reduce some of the infiltration, but generally they result in increasing indoor concentrations as the sources remain constant.

Unlike the outdoor environment, where the vagaries of meteorology and topography can substantially influence both the degree and variability of exposures, concentrations in the indoor environment to which people are exposed are generally more stable and can be determined with reasonable accuracy through measurement techniques. Such techniques include stationary monitors in the indoor environment and personal monitors on individuals to measure specific exposures. Exposures can also be estimated using mathematical models that predict the distribution of pollutants in the indoor environment and include population activity patterns. The available models range widely in complexity and accuracy and generally need to be designed for the specific study.

Finally, assessors in the past often focused on the potential uncertainties in the hazard identification and dose–response steps and did not adequately consider the uncertainties in exposure assessment. That often is inappropriate because exposure assessment uncertainties may be equal to or greater than those associated with the hazard identification and dose–response steps (Patrick 1992).

D. Risk Characterization

The risk characterization step involves bringing together the information obtained in the previous three steps for decision making. As noted by the Commission on Risk Assessment and Risk Management, many risk assessments in the past estimated risks using hypothetical, nonexistent, maximally exposed individuals and they generally neglected frequency, duration, and magnitude of actual population exposures. In addition, they relied on quantitative estimates (often single point estimates) of risk and expected regulatory decision makers to translate that information into appropriate decisions and expected those at risk to understand the implications. The Commission recommended that risk characterizations of the future include information that is useful for all parties in the decision process, and that qualitative information on the nature of the adverse effects and the risk assessment itself should be included with the quantitative estimates of risk. Information on the range of informed views and the evidence supporting them should also be shared.

The EPA provided the most definitive recent guidance on risk characterization in a 1992 memorandum from then Deputy Administrator Henry Habicht to EPA's Assistant Administrators and Regional Administrators entitled, "Guidance on Risk Characterization for Risk Managers and Risk Assessors." The memorandum provided guidance on describing risk assessment results in EPA reports, presentations, and decision packages, and focused on the public perceptions and misperceptions that can occur when confronted with risk information. The memorandum provided guidance intended to do the following:

- present a full and complete picture of risk, including a statement of confidence about the data and methods used to develop the assessment;
- provide a basis for greater consistency and comparability in risk assessments across the Agency programs; and
- ensure that professional scientific judgment plays an important role in the overall statement of risk.

The following sections discuss in more detail how the risk assessment paradigm has been used in the past for carcinogens and noncarcinogens.

V. APPLICATIONS OF THE PARADIGM

A. Carcinogens

The risk assessment paradigm articulated by the National Research Council was precipitated largely by concerns over exposure to environmental carcinogens. The EPA scientists began to grapple with this issue soon after formation of the agency and published the first guidelines for assessing the risks of exposure to carcinogens in the mid-1970s (EPA 1976; Albert et al. 1977). Because the science was in its formative stages, these guidelines contained many assumptions and were generally conservative, meaning that in using the process the risk of cancer would not be underestimated. This is prudent public health policy when there is uncertainty. In the mid-1980s, the EPA published its first detailed guidelines for carcinogen risk assessment (EPA 1986a). These were accompanied by guidelines for mutagenicity risk assessment (EPA 1986e), suspect developmental toxicant risk assessment (EPA 1986c), chemical mixture health risk assessment (EPA 1986d), and exposure assessment (EPA 1986b).

The estimates of risk resulting from exposure to a carcinogen rest on the determination of the cancer potency factor. This factor, which usually represents the risk associated with a unit lifetime average dose or intake level, is multiplied by the average measured or estimated lifetime human intake to estimate risk. The EPA has conducted numerous studies in the past to estimate cancer potency factors, but resource reductions and the sheer number of possible environmental carcinogens means that many outside organizations must now conduct much of the basic research. The EPA maintains a database, the Integrated Risk Information System (IRIS), of scientifically accepted cancer potency factors. Unfortunately, resource limitations have prevented IRIS from being updated as frequently as originally planned.

The evidence that a substance is carcinogenic often comes from many sources (i.e., the workplace, animal tests, and other laboratory studies) and from studies that vary widely in terms of refinement and accuracy. As such, in the 1986 guidelines the EPA developed a system for grading the evidence. The EPA's weight-of-evidence classification contained five categories:

Group A: Human Carcinogens — This category is for substances for which there is clear human evidence (i.e., from epidemiologic studies) supporting a causal association between exposure to the substance and cancer.

Group B: Probable Human Carcinogens — This category is for substances for which there is limited evidence from epidemiology studies of carcinogenicity in humans (Group B1) or for which, lacking adequate evidence in humans, there is sufficient evidence of carcinogenicity in animals (Group B2).

Group C: Possible Human Carcinogens — This category is for substances for which there is limited evidence of carcinogenicity in animals and an absence of data in humans.

Group D: Not Classified — This category is for substances for which there is inadequate evidence for assessing carcinogenicity.

Group E: No Evidence of Carcinogenicity — This category is for substances for which there is no evidence of carcinogenicity in at least two adequate animal tests in different species or in both epidemiologic and animal studies.

Importantly, the weight-of-evidence categorization is qualitative. There is no current scientifically accepted means to assign a quantitative value to the groups. Therefore, cancer risks are calculated using the cancer potency factor and exposure assessment results and the weight-of-evidence is normally expressed along with the calculated risk value.

More recently, the EPA proposed revised guidelines for carcinogen risk assessment (EPA 1996). These guidelines take a much more direct, narrative approach to weighing evidence for carcinogenic hazard potential. Group A substances are designated as "known/likely" carcinogens, Group E substances are designated as "not likely" to be carcinogens, and everything in between is designated as "cannot be determined." These changes reflect the difficulties the Agency had in making regulatory decisions using the original weight-of-evidence classifications.

Both individual and total population risks may be calculated for carcinogens. The individual risk is the cancer risk estimated to be experienced by an individual from a lifetime of exposure at a specified potency and exposure. Lifetime often was assumed in the past to be the average U.S. human life span of 70 years, but more recently is assumed to be the likely time of residence near a U.S. source or the lifetime of most U.S. industrial facilities, both 30 years. The most commonly estimated risk in the past was the risk to the maximally exposed individual (MEI). However, the use of the MEI has many critics because it often is based on unrealistic conditions such as a person living outdoors for 70 years at the fenceline of the source emitting the pollutant. The EPA is currently moving away from using the term MEI. In its report *Science and Judgment in Risk Assessment* (NRC 1994), the National Research Council states that "EPA no longer uses the term MEI, noting the difficulty in estimating it and the variety of its uses. The MEI has been replaced with two other estimators of the upper end of the individual exposure distribution, a 'high-end exposure estimate' (HEEE) and the theoretical upper-bounding estimate (TUBE)." The EPA's Exposure Assessment Guidelines (EPA 1992) define HEEE as a "plausible estimate of the individual exposure of those persons at the upper end of the exposure distribution." High-end is stated conceptually as "above the 90th percentile of the population distribution, but not higher than the individual in the population who has the highest exposure." The TUBE is defined in the Guidelines as "a bounding calculation that can easily be calculated and is designed to estimate exposure, dose, and risk levels that are expected to exceed the levels experienced by all individuals in the actual distribution. The TUBE is calculated by assuming limits for all variables used to calculate exposure and dose that, when combined, will result in the mathematically highest exposure or dose."

Population risks are generally calculated by evaluating the distribution of individual risk across the exposed population (e.g., the number of individuals at risk in various risk intervals, such as 10^{-4}, 10^{-5}, and 10^{-6}) and estimated average annual incidence. Estimating exposures for all of the population exposed to an outdoor pollutant or pollutants of concern can encompass many miles surrounding the source. Mathematical models are widely used to simulate the conditions giving rise to the potential risk. Although the accuracy of the models varies widely, some accepted dispersion models are validated to about 30 miles from a source with uncomplicated terrain and meteorology. Estimating exposures for indoor populations is generally easier, often being determined directly through measurement and population activity studies.

A risk distribution (associated with either outdoor or indoor exposures) might be calculated and could look like the following:

Risk range	Population
Above 10^{-1}	1
10^{-1} to 10^{-2}	10
10^{-2} to 10^{-3}	100
10^{-3} to 10^{-4}	1,000
10^{-4} to 10^{-5}	10,000
10^{-5} to 10^{-6}	100,000
Below 10^{-6}	1,000,000

In general, the greater the number of people at high risk levels, the greater the likelihood that risk distribution will play a role in the regulatory decision making.

Another population risk is average annual incidence. In this calculation, the number of people at a specific risk range is multiplied by that risk and the number of cancer deaths is estimated. For example, if ten people are exposed to a carcinogen at a risk level of one in ten, one cancer death is estimated. In the example above, each range results in one estimated cancer death, so the total is seven. Since the cancer risks are for a 70-year lifetime, the average annual incidence is 0.1 cancer death.

Because cancer is one of the most widespread illnesses (i.e., about one in four Americans will contract cancer of some form in their lifetime and about one in five will die prematurely from cancer), environmental cancer risks are often compared with the national average. This almost always shows environmental cancer risks associated with exposure to a specific pollutant of concern as approaching insignificance in comparison with the national average total cancer rate. Such comparisons are inappropriate because there are literally hundreds of different cancers in humans resulting from genetic, lifestyle, natural, and environmental causes, and humans typically are exposed simultaneously to many potential carcinogens.

Cancer risk is usually calculated by multiplying the chronic daily intake by the cancer potency. The chronic daily intake is calculated using the following type of equation (this one addresses air exposures) (EPA 1989):

$$CDI = \frac{CA \times IR \times ED \times EF \times L}{BW \times AT \times 365}$$

where:

CDI = chronic daily intake (mg/kg/day)
 CA = contaminant concentration (mg/m^3)
  IR = inhalation rate (m^3/hour)
 ED = exposure duration (hours/week)
 EF = exposure frequency (weeks/year)
   L = length of exposure (years)
BW = body weight (kg)
 AT = averaging time (period over which exposure is averaged — usually 70 years for carcinogens)
 365 = days per year

The cancer potency is usually expressed as the reciprocal of mg/kg/day to achieve a dimensionless risk value. For carcinogens, the cancer risk is often expressed numerically as a probability (e.g., one chance in one million). For simultaneous exposures to several carcinogens, public health officials generally assume that cancer risks are additive even though there may be no direct relationship.

There is no cancer risk posed by a single substance that is widely accepted as insignificant, although a cancer risk of one in one million (1×10^{-6}) resulting from exposure to an individual substance often is postulated as a *de minimis* risk. Some EPA programs in the past, based on regulatory policy rather than legislative direction, have used risk criteria ranging from one in ten thousand (1×10^{-4}) to one in ten million (1×10^{-7}). In response to the 1986 vinyl chloride ruling by the U.S. Court of Appeals, in 1989 the EPA promulgated acceptable risk guidelines for hazardous air pollutants that specified the following:

• protection of the greatest number of persons to an individual lifetime risk of no more than one in one million (1×10^{-6}), and
• limiting to no higher than one in ten thousand (1×10^{-4}) the risk to the person exposed to the maximum lifetime concentration.

That decision was supported by Congress in the 1990 Clean Air Act Amendments (see section 112(f)(2)(B)) but was not established as law. In general, the acceptability of cancer risks should not be a scientific choice but, rather, a policy choice made by public health officials in a public process.

New methods of evaluating cancer risks are being proposed and developed. Ideally, the modeling should reflect the underlying mechanisms. For example, the

MKV model (Moolgavkar and Knudson 1981) includes consideration of cell turn-over rates and other nongenotoxic events.

B. Noncarcinogens

For substances associated with a presumed threshold of effect, the typical risk calculation process is more straightforward. The basic process has been used for many years, for example, in setting safe workplace exposure levels. As described by Faustman in the Commission on Risk Assessment and Risk Management report (1997), the process rests on the calculation of a factor often called the risk reference dose (RfD), risk reference concentration (RfC), or acceptable daily intake (ADI). These factors are usually derived by identifying no-observed-adverse-effect-levels (NOAELs) and dividing them by uncertainty or modifying factors. More recently, benchmark doses (BMDs) are being developed using curve-fitting procedures to find a dose that produces a specific effect. Confidence limits are then generated around that dose, which is set at a lower confidence limit to produce a specified percentage change in response. The BMD is then used to calculate an RfC or RfD.

The RfD is defined by the EPA as an estimate of the daily dosage to a substance that is likely to be without an appreciable risk of deleterious effect during the lifetime of exposure to a human (EPA 1986f). As used by the EPA, the purpose of the RfD is to provide a benchmark with which other route-specific doses (e.g., specific human exposures) can be compared. Doses that are less than the RfD are usually believed to be of no concern; doses that are greater than the RfD are generally viewed as indicating an increased probability of an adverse effect and the need for further study. In developing the RfD approach, the EPA stated that the RfD is an approximate number with an uncertainty spanning perhaps an order of magnitude (EPA 1986f). Although doses higher than the RfD have a higher probability of producing an adverse effect, the RfD is generally not viewed by the EPA as an "unacceptable" risk level *per se* but, rather, a generally conservative estimate of a maximum acceptable dose.

RfDs are normally derived from workplace exposures or from animal studies and, thus, rest on the assumptions necessary to extrapolate results from animals to man and from high to low doses. These extrapolations are typically made by using uncertainty factors. For example, an uncertainty factor of ten may be added to account for the use of animal test results, another uncertainty factor of ten may be added to account for the use of test data from another exposure route, and more precise factors may be added to compensate for different lengths of exposure. The type and level of the uncertainty factors to be used in any analysis must be selected with great care by those thoroughly familiar with toxicological analyses.

Noncarcinogenic risks are typically estimated by comparing the RfD with the maximum daily intake (MDI). The MDI is derived using the following equation:

$$MDI = \frac{CA \times IR \times 24}{BW}$$

where:

MDI = maximum daily intake (mg/kg-day)
CA = contaminant concentration (mg/m^3)
IR = inhalation rate (m^3/hour)
24 = hours per day
BW = body weight (kg)

Risk is inferred by comparing the MDI with the RfD. As noted earlier, an MDI that is greater than the RfD is viewed as indicating an increased probability of an adverse effect and the need for further study. This comparison is usually made by dividing the MDI by the RfD. The result is called the Hazard Quotient. Again, when exposure to more than one substance is being evaluated, the individual hazard quotients are added and the result is usually called the Hazard Index. As with the evaluation of combined effects of mixtures of carcinogens, this process for non-carcinogens is conservative. The process for evaluating noncarcinogenic mixtures is described in detail in the EPA's guideline for assessment of chemical mixtures (EPA 1986d).

The acceptability of noncancer risks is established from the hazard quotient and hazard index calculations. In general, if the quotient or index is below one, it is acceptable; if it is above one, it may be unacceptable depending upon how far above and upon the uncertainties in the calculation process. As noted earlier, the EPA generally does not consider the RfD a clear pass-fail criterion. Again, the accept-ability of noncancer risks is a not a scientific decision but rather a policy choice that should be made by public health officials in a public process.

While the discussion here focuses on current practices of the EPA, which addresses risk assessment more directly than any other organization at this time, several other U.S. and international organizations have established noncarcinogenic guidelines or criteria levels. These include OSHA, the Department of Defense (DOD), and the American Industrial Hygiene Association (AIHA) in the U.S. The DOD developed exposure limits for military personnel operating in emergency situations. The DOD limits include emergency exposure guidance levels (EEGL), short-term public guidance levels (SPEGL), and continuous exposure guidance levels (CEGL). The AIHA limits are known as emergency response planning guidelines (ERPG) and were developed as a result of the chemical accident in Bhopal, India, in 1984.

Newer methods are being developed to assist in the assessment of noncancer health effects. For example, a method referred to as the decision analytic approach (Richmond 1991) utilizes expert judgment in addressing key uncertainties. The view is that scientific data alone will rarely be sufficient for decision making and that probabilistic exposure–response relationships can be elicited from knowledgeable health scientists to supplement the scientific data. These approaches have been used in recent years by the EPA to assist in decisions on two criteria air pollutants, ozone and lead.

VI. RISK MANAGEMENT

Risk management is the process of making appropriate decisions using the information produced by the risk assessment. As noted earlier, the National Research Council (NRC 1983) first crystallized the concepts of risk assessment and *risk management* for federal government environmental decision making and defined the term risk management as the complex of judgment and analysis that uses the results of risk assessment to produce a decision about an environmental action. In this context, risk management generally includes: (1) developing regulatory options; (2) evaluating the public health, economic, social, and political consequences of the regulatory options; (3) making regulatory decisions; and (4) taking regulatory actions. The EPA (EPA 1984) describes the two major uses of the risk management approach: (1) setting priorities among risks in the environment that are amenable to control by the EPA, and (2) choosing the appropriate reduction actions for the risks so selected. In general, the balancing that goes into such risk management decisions includes consideration of the harmful effects, the costs, and the confidence in the decision. The EPA also notes that while individual risk management decisions may be perceived as balancing risk reduction against resources, the system as a whole was designed to balance risk against risk.

More recently, the Commission on Risk Assessment and Risk Management (Commission 1997) recommended a move away from the chemical-by-chemical, medium-by-medium, risk-by-risk strategy of the past. That strategy evolved from multiple, unregulated statutory requirements and produced many effective risk management decisions; however, a more integrated, effective environmental management program requires a risk management framework that can engage a range of stakeholders and address the interdependence and cumulative effects of various problems. The Commission stated that the framework must have the capacity to address various media, contaminants, and sources of exposure, as well as an array of public values, perceptions, and ethics. Importantly, there is a need to address multiple chemical exposures such as those experienced in the indoor environment. The ideal risk management framework, as described by the Commission, includes the following stages:

Problem/Context — The problem is identified and examined in a comprehensive, public-health context. Stakeholders are identified at this stage and included thereafter.

Risks — Risks are determined by considering the nature, likelihood, and severity of the adverse effects. Risks are evaluated by scientists with input from the stakeholders. The factual and scientific basis of the problem is articulated and incorporated into a characterization of the risks.

Options — A variety of approaches to addressing the problem are identified by scientists, regulators, and stakeholders. Ideally, both regulatory and nonregulatory approaches are identified. Implementation considerations are also identified, which might include financial, political, legal, and cultural factors.

Decisions — Ideally, the most feasible, effective, acceptable, and cost-effective approaches to mitigating the problem will be identified in cooperation with the affected and responsible parties. In some instances, legislative or regulatory requirements will supersede this process or a consensus cannot be reached. In the end, the responsible regulatory authority makes the necessary decision.

Actions — Necessary actions are taken to implement the decisions in a public process and changes are made if necessary.

Evaluation — There is often insufficient follow-up after an action is taken to evaluate the effectiveness and cost of the action, or to compare the results with the estimates made in the earlier stages. Appropriate risk management dictates this follow-up through monitoring or surveillance, discussion with affected parties, and analysis of health or environmental indicators. On the basis of this evaluation, the original problem might need to be redefined, the actions reconsidered, and the process repeated.

BIBLIOGRAPHY

Albert, R.E., Train, R.E., et al. 1977. Rationale developed by the U.S. Environmental Protection Agency for the assessment of carcinogenic risk, *Journal of the National Cancer Institute* 58:1537.

American Industrial Hygiene Association. 1994. Comments on OSHA's Proposed Indoor Air Quality Standard, *59 FR 15968-16039.*

Commission on Risk Assessment and Risk Management. 1997. *Risk Assessment and Risk Management in Regulatory Decision Making.*

Environmental Protection Agency (EPA). 1976. Interim procedures and guidelines for health risks and economic impact assessments of suspected carcinogens, *41 FR 21402.*

Environmental Protection Agency (EPA). 1984a. Approaches to Risk Assessment for Multiple Chemical Exposures, Report No. EPA-600/9-84-008, Environmental Criteria and Assessment Office, U.S. Environmental Protection Agency.

Environmental Protection Agency (EPA). 1984b. Risk Assessment and Management: Framework for Decision Making, Report No. EPA 600/9-85-002, U.S. Environmental Protection Agency, December, 1984.

Environmental Protection Agency (EPA). 1986a. *Integrated Risk Information System (IRIS) Database, Appendix A, Reference Dose (RfD): Description and Use in Health Risk Assessments,* Office of Health and Environmental Assessment, U.S. Environmental Protection Agency, Washington, DC.

Environmental Protection Agency (EPA). 1986b. Part II. Guidelines for carcinogen risk assessment, *51 FR 33992.*

Environmental Protection Agency (EPA). 1986c. Part III. Guidelines for mutagenicity risk assessment, *51 FR 34006.*

Environmental Protection Agency (EPA). 1986d. Part IV. Guidelines for health risk assessment of chemical mixtures, *51 FR 34014.*

Environmental Protection Agency (EPA). 1986e. Part V. Guidelines for health assessment of suspect developmental toxicants, *51 FR 34028.*

Environmental Protection Agency (EPA). 1986f. Part VI. Guidelines for exposure assessment, *51 FR 34042.*

Environmental Protection Agency (EPA). 1989. Risk Assessment Guidance for Superfund, Volume 1, Human Health Evaluation Manual (Part A), Report No. EPA 540/1-89-002, Office of Solid Waste and Emergency Response, U.S. Environmental Protection Agency, Washington, DC.

Environmental Protection Agency (EPA). 1992. Guidelines for exposure assessment, *57 FR 22888*.

Environmental Protection Agency (EPA). 1996. Proposed Guidelines for Carcinogen Risk Assessment, Report No. EPA/600/P-92/003C, U.S. Environmental Protection Agency, April 1996.

Lipfert, F.W. 1994. *Air Pollution and Community Health: A Critical Review and Data Sourcebook*, Van Nostrand Reinhold: New York.

Moolgavkar, S.H., Knudson, A.G. 1981. Mutation and cancer: A model for human carcinogenesis, *Journal of the National Cancer Institute* 66:1037.

Moschandreas, D.J. 1988. Risk assessment and the indoor air environment, *State of the Art Review (Healthy Buildings '88 Conference, Stockholm, Sweden)* 1:19–23.

National Research Council (NRC). 1983. *Risk Assessment in the Federal Government: Managing the Process*, prepared by the Committee on Institutional Means for Assessment of Risk to Public Health, Commission on Life Sciences, National Academy Press: Washington, DC.

National Research Council (NRC). 1994. *Science and Judgment in Risk Assessment*, Prepared by the Committee on Risk Assessment of Hazardous Air Pollutants, Commission on Life Sciences, National Academy Press: Washington, DC.

Özkaynak, H., Spengler, J.D. 1990. Introduction to the risk assessment workshop on indoor air quality, *Risk Analysis* 10(1):15.

Patrick, D.R. 1992. *The Impact of Exposure Assessment Assumptions and Procedures on Estimates of Risk Associated with Exposure to Toxic Air Pollutants*, Paper No. 92-95.02, Presented at the 85th Annual Meeting of the Air and Waste Management Association, Kansas City, MO, June, 1992.

Richmond, H.M. 1991. *Overview of a Decision Analytic Approach to Noncancer Health Risk Assessment*, Paper No. 91-173.1, Presented at the 84th Annual Meeting of the Air and Waste Management Association, Vancouver, BC, June, 1991.

Ruckelshaus, W.D. 1983. Statement before the Subcommittee on Oversight and Investigations, Committee on Energy and Commerce, U.S. House of Representatives, November 7, 1983.

CHAPTER **3**

Hazard Identification
of Indoor Air Pollutants

John J. Liccione

CONTENTS

1-56670-323-9/99/$0.00+$.50
© 1999 by CRC Press LLC

I. INTRODUCTION

The term *hazard identification* is widely used in risk assessment. The framework for hazard identification was provided by the National Research Council (NRC) in their seminal 1983 risk assessment guidelines, in which hazard identification was defined as "the process of determining whether exposure to an agent causes an increase in the incidence of a health condition (e.g., birth defects, cancer)" (NRC 1983). Hazard identification is the first step of the risk assessment process and entails the characterization of the nature and strength of the evidence of causation. The focus of hazard identification is on answering the question, "Does the agent cause the adverse effect?"

The NRC guidelines also identified four general classes of information that may be used in the hazard identification step, including: (1) epidemiological data, (2) animal-bioassay data, (3) short-term studies, and (4) comparisons of molecular structure. Each of these classes is further characterized by a number of components, as depicted in NRC 1983, and summarized in Table 3.1.

The essential features of hazard identification as outlined by the NRC were subsequently adopted by the U.S. Environmental Protection Agency (EPA). The EPA subsequently established risk assessment guidelines for carcinogens (EPA 1986a), mutagens (EPA 1986b), reproductive toxins (EPA 1996b), neurotoxins (EPA 1995a), and developmental toxins (EPA 1986c; 1991a). Recently, the EPA published important proposed revisions to the guidelines for carcinogens (EPA 1996). In addition, at the time this book was written in 1997, guidelines for immunotoxicity were being developed by the EPA. In all of these EPA guidelines, the concept of hazard identification consists of two important components:

1. The identification of a potential hazard, and
2. The assignment of a "weight of evidence" describing the strength of the information bearing on the potential for a particular hazard.

Hazard identification also entails the quantification of the concentration of a particular contaminant at which it is present in the environment.

Originally, hazard identification was used primarily to identify the potential hazards of chemicals in ambient air, food, and water. In recent years, there has been growing concern over the health hazards of indoor air pollutants. This chapter illustrates the application of the hazard identification process to the study of indoor air pollutants. Additionally, the limitations and difficulties related to the interpretation of data obtained from the application of hazard identification in this arena are addressed.

II. APPROACHES TO THE HAZARD IDENTIFICATION OF INDOOR AIR POLLUTANTS

A wide variety of health effects have been attributed to exposure to indoor air pollutants. The primary potential health effects include acute and chronic respiratory

Table 3.1   Information Used in Hazard Identification

Classes of Information	Components
Epidemiologic Data	What relative weights should be given to studies with differing results? For example, should positive results outweigh negative results if the studies that yield them are comparable? Should a study be weighted in accord with its statistical power?
	What relative weights should be given to results of differing types of epidemiologic studies? For example, should the findings of a prospective study supersede those of a case-control study, or those of a case-control study supersede those of an ecologic study?
	What statistical significance should be required for results to be considered positive?
	Does a study have special characteristics (such as the questionable appropriateness of the control group) that lead one to question the validity of its results?
	What is the significance of a positive finding in a study in which the route of exposure is different from that of a population at potential risk?
	Should evidence about different types of responses be weighted or combined (e.g., data on different tumor sites and data on benign versus malignant tumors)?
Animal-Bioassay Data	What degree of confirmation of positive results should be necessary? Is a positive result from a single animal study sufficient, or should positive results from two or more animal studies be required? Should negative results be disregarded or given less weight?
	Should a study be weighted according to its quality and statistical power?
	How should evidence of different metabolic pathways or vastly different metabolic rates between animals and humans be factored into a risk assessment?
	How should the occurrence of rare tumors be treated? Should the appearance of rare tumors in a treated group be considered evidence of carcinogenicity even if the finding is not statistically significant?
	How should experimental-animal data be used when the exposure routes in experimental animals and humans are different?
	Should a dose-related increase in tumors be discounted when the tumors in question have high or extremely variable spontaneous rates?
	What statistical significance should be required for results to be considered positive?
	Does an experiment have special characteristics (e.g., the presence of carcinogenic contaminants in the test substance) that lead one to question the validity of its results?
	How should findings of tissue damage or other toxic effects be used in the interpretation of tumor data? Should evidence that tumors may have resulted from these effects be taken to mean that they would not be expected to occur at lower doses?
	Should benign and malignant lesions be counted equally?
	Into what categories should tumors be grouped for statistical purposes?
	Should only increases in the numbers of tumors be considered, or should a decrease in the latent period for tumor occurrence also be used as evidence of carcinogenicity?

(continues)

Table 3.1   (continued)

Classes of Information	Components
Short-Term Test Data	How much weight should be placed on the results of various short-term tests?
	What degree of confidence do short-term tests add to the results of animal bioassays in the evaluation of carcinogenic risks for humans?
	Should *in vitro* transformation tests be accorded more weight than bacterial mutagenicity tests in seeking evidence of a possible carcinogenic effect?
	What statistical significance should be required for results to be considered positive?
	How should different results of comparable tests be weighted? Should positive results be accorded greater weight than negative results?
Structural Similarity to Known Carcinogens	What additional weight does structural similarity add to the results of animal bioassays in the evaluation of carcinogenic risks for humans?
General	What is the overall weight of the evidence of carcinogenicity? (This determination must include a judgment of the *quality* of the data presented in the preceding section.)

Source: NRC 1983.

effects, neurological toxicity, lung cancer, eye and throat irritation, reproductive effects, and developmental toxicity. In some instances, odor may reveal the presence of a potential hazard; however, odor is not always reliable, especially for the identification of potential long-term exposures to low concentrations of an indoor air pollutant.

Adverse health effects can be useful indicators of an indoor air quality problem (EPA 1995b). The approaches that may be used to gain evidence that a suspect indoor air pollutant causes a specific adverse health effect are discussed in more detail below.

A. Neurotoxicity

Fatigue, headaches, dizziness, nausea, lethargy, and depression are classic neurological symptoms that have been associated with indoor air pollutants. The EPA risk assessment guidelines for neurotoxicity (EPA 1995a) address hazard identification as it pertains to the neurotoxicity of chemicals in general. Based on these guidelines, the hazard identification of a potential neurotoxin "involves examining all available experimental animal and human data and the associated doses, routes, timing, and durations of exposure to determine if an agent causes neurotoxicity in that species and under what conditions." Moreover, the guidelines provide guidance on how to interpret data relating to various neurological endpoints, including structural endpoints, neurophysiological parameters (e.g., nerve conduction and electro-encephalography), neurochemical changes (e.g., neurotransmitter levels), behavioral effects (e.g., functional observation battery), and developmental neurotoxic effects.

Other considerations include interpretation of pharmacokinetic data, comparisons of molecular structure, statistical factors, and *in vitro* neurotoxicity data.

An approach that may have significant utility for the specific identification of potential neurotoxic indoor air pollutants was described by Otto and Hudnell (1993). This approach involves the application of visual evoked potentials (VEP) and chemosensory evoked potentials (CSEP) in the evaluation of the effects of acute and chronic chemical exposure. The similarity of VEP waveforms in different species renders this feature useful for cross-species extrapolation. Numerous chemicals, including solvents, metals, and pesticides (many of which have been confirmed as indoor air pollutants), were reported to alter VEP in humans and/or animals.

Otto and Hudnell also discuss the methodology that can be used to elicit various VEPs (e.g., flash evoked potentials by stroboscopic presentation of a diffuse flashing light, pattern-reversal VEPs by a reversing checkerboard pattern, and sine-wave grating VEPs by sinusoidal gratings). The advantages and disadvantages of each type of VEP are discussed, and stimulus patterns associated with each are illustrated. In addition, VEPs have been applied to detect subtle subclinical signs of polyneuropathy in workers exposed to solvents. One kind of VEP, flash evoked potentials (FEP), has been used to evaluate impaired visual function in workers exposed to solvents such as n-hexane and xylene. Pesticides, metals, anesthetics, and gases also have been found to alter FEPs.

CSEPs represent a type of evoked potential that may be useful for an objective measurement of chemosensory response. Measurement of chemosensory function is relevant to the hazard identification of indoor air pollutants because odors and sensory irritation of the eyes, nose, and throat provide vital and early warning signs of a potential hazard. Trigeminal somatosensory evoked potentials have been shown to provide a reliable method to detect trigeminal lesions in workers as the result of long-term exposure to the solvent trichloroethylene. Otto and Hudnell provide a description of CSEPs waveforms, the effects of habituation on the evoked potential, and how to distinguish olfactory from trigeminal CSEPs. CSEPs recorded in conjunction with psychophysical or rating scale measures of sensory irritation could be used to evaluate objectively the effects of volatile organic compounds, to distinguish between olfactory and trigeminal components of sick building syndrome, and to assess the reported hypersensitivity of multiple chemical sensitivity patients to chemicals.

Sram et al. (1996) describe the use of the Neurobehavioral Evaluation System (NES2) in the assessment of the impacts of air pollutants on sensorimotor and cognitive function in children. The NES2 is a computerized assessment battery that is ideal for neurotoxicity field testing. It consists of tests for finger tapping, visual digit span, continuous performance, symbol-digit substitution, pattern comparison, hand-eye coordination, switching attention, and vocabulary.

B. Carcinogenicity

Several indoor air pollutants have been implicated in the risk of cancer, in particular, lung cancer. The 1986 EPA cancer risk assessment guidelines provide an approach to the hazard identification of potential carcinogens (EPA 1986a). These guidelines discuss how to derive a weight-of-evidence for carcinogenicity on the basis of data from epidemiologic and animal toxicity studies, genotoxicity studies, and structure-activity relationships. Both malignant and benign tumors are considered in the evaluation of carcinogenic hazard. The concept of the significance of the maximum tolerated dose (MTD) in the design of animal carcinogenicity bioassays is discussed.

As described more fully in Chapter 2, the EPA 1986 cancer risk assessment guidelines originally established the following classification scheme for carcinogens:

Group A — Human Carcinogens
Group B — Probable Human Carcinogens
Group C — Possible Human Carcinogens
Group D — Not Classified
Group E — No Evidence of Carcinogenicity

The International Agency for Research on Cancer (IARC) has developed a similar ranking scheme.

The EPA's cancer guidelines also state that the weight-of-evidence that an agent is potentially carcinogenic for humans increases under the following conditions:

- with the increase in number of tissue sites affected by the agent;
- with the increase in number of animal species, strains, sexes, and number of experiments and doses showing a carcinogenic response;
- with the occurrence of clear-cut dose–response relationships as well as a high level of statistical significance of the increased tumor incidence in treated compared to control groups;
- when there is a dose-related shortening of the time-to-tumor occurrence or time to death with tumor; and
- when there is a dose-related increase in the proportion of tumors that are malignant.

More recently, the EPA revised and extended the 1986 guidelines in new draft proposed guidelines (EPA 1996a). A noteworthy change in these new proposed cancer guidelines is the incorporation of mechanistic and pharmacokinetic data into the hazard identification of carcinogens. The guidelines also discuss the significance of threshold versus nonthreshold mechanisms, and address the relevancy of certain tumor types in animals (e.g., renal tumors associated with hyaline droplet nephropathy) to humans. The proposed cancer guidelines provide a less structured classification of human carcinogenic potential, grouping substances only in the classifications "known/likely carcinogen," "cannot be determined," and "not likely."

Genotoxicity data can provide insight into the mechanism of carcinogenicity (e.g., nongenotoxic versus genotoxic carcinogen). Short-term genetic bioassays have been applied to the study of potential mutagenic indoor air pollutants (Lewtas et al. 1993). The standard Salmonella forward mutation assay and the Salmonella reverse mutation assay, in particular, have been useful. Since the first bioassay studies of indoor air pollutants required the collection of large volumes of air, modifications have been made to the standard mutagenicity assays so that smaller volumes can be tested. These modified assays have been termed *microsuspension mutagenicity assays*. Combined with improved sampling techniques (e.g., special exposure chambers, the use of filters and electrostatic precipitators, and extraction by ultrasonication), these assays allow for the examination of the genotoxic potential of complex mixtures of indoor air pollutants. Results of various studies have revealed that environmental tobacco smoke (ETS) is the major source of mutagens indoors (Lewtas et al. 1993).

C. Respiratory and Sensory Irritative Effects

Respiratory effects are common complaints that have been linked to exposure to indoor air pollutants. These effects include irritation, inflammation, wheezing, cough, chest tightness, dyspnea, respiratory infections, lung function decrement, respiratory hypersensitivity, acute respiratory illness, and chronic respiratory diseases (Samet and Speizer 1993; Becher et al. 1996). A variety of methods has been used in epidemiologic and controlled chamber human studies to assess the potential respiratory and irritative effects of indoor air pollutants. Some of the more common methods employed in human studies are discussed in more detail in the following paragraphs.

The American Thoracic Society established guidelines with a rather high degree of standardization on pulmonary function testing and respiratory symptom questionnaires (IARC 1993). Respiratory symptom questionnaires are particularly sensitive for assessing chronic symptoms like cough, sputum production, wheezing, and dyspnea (Samet and Speizer 1993).

Spirometry has been the most widely used technique for the measurement of pulmonary function in human studies (Samet and Speizer 1993). This technique involves the collection of exhaled air during the forced vital capacity maneuver, and allows for the determination of forced vital capacity (FVC), the total amount of exhaled air, and the volume of air exhaled in the first second (FVC_1). It also permits measurements of flow rates at lower lung volumes, indications of an adverse effect on the small airways of the lung. Small airway dysfunction can also be assessed by nitrogen washout curves, a possible marker for early toxicity to the lung (IARC 1993).

Hypersensitivity and nonspecific hyperreactivity are parameters less frequently examined in human studies (IARC 1993). However, methods such as histamine or methacholine challenge for nonspecific hyperreactivity and skin allergen tests for hypersensitivity can be utilized (IARC 1993; Samet and Speizer 1993).

D. Immunological Effects

There is concern for the potential immunological effects of indoor air pollutants. A number of health effects, such as respiratory hypersensitivity associated with exposures to indoor air pollutants, may involve immunological mechanisms (Vogt 1991; Chapman et al. 1995). Immunochemical and molecular methods for defining and measuring indoor allergens are available (Chapman et al. 1995). Studies have also shown that IgE-mediated sensitization to indoor allergens (e.g., dust mite and fungi) can cause asthma, and may play some role in the development of perennial rhinitis and atopic dermatitis (Chapman et al. 1995).

Indoor allergens can now be detected by monoclonal and polyclonal antibody based, enzyme-linked immunosorbent assay (ELISA) techniques (Chapman et al. 1995; Burge 1995). For instance, two-site ELISA immunoassays have been used for the characterization of dust mite, animal dander, cockroaches, and *aspergillus* (Burge 1995). Epidemiological studies employing standardized sampling techniques and extraction procedures have allowed for the determination of risk levels of exposure for the development of IgE sensitization (e.g., 2 µg dust mite/g dust) and determination of threshold levels for the development of allergic symptoms (e.g., 10 µg dust mite/g dust) (Chapman et al. 1995).

Besides ELISA methods, other immunoassay techniques are available for detecting the presence of specific indoor air allergens (Burge 1995). One such method is the radioallergosorbent test (RAST) for measuring allergen-specific IgE antibodies. Inhibition of antibody binding on immunoblots ("immunoprint inhibition") is another method. Finally, chemical assays as indicators of allergen sources (e.g., the guanine assay for dust mites) have been described.

Immunological biomarkers may have utility for the identification of health hazards arising from exposure to indoor air pollutants (Vogt 1991). Vogt also discusses immune biomarkers that may be useful for identifying potential immunotoxic indoor air pollutants; these include the following:

- tests for antigen-specific IgE antibodies (skin testing or *in vitro* assays);
- assays for auto-antibodies;
- tests for humoral mediators, e.g., the serum proteins involved with inflammatory responses (such as complement) may provide some indication of irritative or immune reactions to air pollutants;
- analysis of peripheral blood leukocytes and lymphocytes; and
- examining immune cells from accessible mucosal surfaces such as nasal scrapings; this was described as the most promising approach to cellular assessment for indoor air exposures.

E. Developmental and Reproductive Effects

Several chemicals that have been detected in the indoor environment are considered potential developmental and or reproductive toxins. Hazard identification as applied to the developmental and reproductive toxicity was addressed by the EPA's Office of Pesticide Programs (EPA 1991a; EPA 1996b). These risk assessment guidelines outline important considerations when using all available studies for hazard identification, namely: (1) reproducibility of results, (2) the number of species affected, (3) pharmacokinetic data, structure activity relationships, and other toxicological data, (4) the number of animals examined in a study, (5) how well a study is designed, (6) consistency in the pattern of developmental or reproductive effects, and (7) maternal toxicity for developmental studies.

The EPA's Office of Prevention, Pesticides and Toxic Substances (OPPTS) also developed harmonized test guidelines that provide guidance on developmental toxicity and reproductive toxicity testing in animals. In addition, the guidelines are designed to ensure that studies are uniformly performed and that information concerning the developmental or reproductive effects of exposure are adequately reported. The guidance includes appropriate methodology, choice of species, endpoints to be examined, and interpretation of the results.

The harmonized developmental guidelines consider important aspects of developmental toxicity such as preliminary toxicity screening, inhalation toxicity testing, and prenatal toxicity. The developmental guidelines also discuss the importance of determining whether developmental toxicity, either reversible or irreversible, has occurred and if it is unrelated to maternal toxicity. The focus of the harmonized reproductive and fertility guidelines is on the design and conduct of a two-generation reproduction study.

The potential developmental and reproductive effects of air pollution can be assessed in epidemiologic studies. For instance, as part of the Teplice Program to investigate the impact of air pollution on the health of the population in the district of Teplice, Czech Republic, low birth weight, congenital malformations, premature births, and fetal loss were examined in a prospective cohort design (Sram et al. 1996). For the reproductive portion of the study, a comparison of reproductive health and semen quality outcomes in males living in Teplice with those of males living in another area was performed.

III. HAZARDS OF SPECIFIC INDOOR AIR CONTAMINANTS

A diversity of pollutants has been detected in indoor air environments. Table 1 in Chapter 1 summarizes the primary indoor air pollutants. This section reviews the health hazards that have been attributed to select indoor air pollutants, specifically, particulates, chemicals including pesticides, volatile organics, combustion products, tobacco smoke, and biological contaminants. Since there are extensive reviews on some indoor air pollutants such as lead and radon, these will not be discussed in any detail.

A. Particulates

The adverse health hazards of ambient levels of particulate matter have been known for quite some time (Dockery and Pope 1994). In particular, increased morbidity and mortality associated with acute episodes of air pollution during the 1930s, 1940s, and 1950s in Meuse Valley, Belgium, Donora, Pennsylvania, and London, England are well documented, although the adverse effects cannot be solely attributed to particulate matter. Other effects attributed to acute exposure to particulate matter are asthma, lung function changes, cough, sore throat, chest discomfort, sinusitis, and nasal congestion. Epidemiological studies suggest chronic respiratory diseases and symptoms, and increased mortality following long-term exposure to respirable particulate air pollution (Pope et al. 1995).

Early investigators quickly recognized that particulate matter is also an indoor air pollutant. Moreover, concentrations of indoor particulate matter can be quite different from outdoor levels. Consequently, studies typically determine outdoor and indoor relationships of particulate matter. It has been difficult, however, to fully separate the effects of indoor particulates from outdoor particulates.

B. Chemicals

1. Pesticides

Pesticides are a large class of compounds that includes organophosphates, carbamates, dicoumarins, and chlorinated hydrocarbons (Cooke 1991). Pesticides are used in the indoor environment as insecticides, rodenticides, germicides, and termiticides in the control of insects, fungi, bacteria, and rodents. In a pilot study, the EPA detected 22 diverse pesticides in the indoor air of homes, 17 of which were detected in the breath of occupants. Monitoring data revealed that the five most prevalent pesticides were chloropyrifos, diazinon, chlordane, propoxur, and heptachlor. Besides direct indoor application, indoor concentrations of pesticides may originate from other sources such as pesticides applied outdoors that then become airborne, or from pesticides that are carried indoors attached to foodstuffs or in the water supply.

Short-term exposure to high concentrations of well-known pesticides, such as heptachlor, aldrin, chlordane, and dieldrin, may result in headaches, dizziness, muscle twitching, weakness, tingling sensations, and nausea (EPA 1995b). Long-term exposure may cause liver and central nervous system effects, as well as increased cancer risk (EPA 1995b).

2. VOCs

Volatile organic compounds (VOCs) represent a large and diverse class of chemicals that possess the ability to volatilize into the atmosphere at normal room temperature (Samet et al. 1988; Cooke 1991). VOCs have been linked to the development of sick building syndrome (Kostiainen 1995); however, the cause of this syndrome is still unclear. Many of the VOCs that have been detected indoors are neurotoxic (Cooke 1991). Clinical signs of VOCs consist of headache, nausea, irritation of the eyes, mucous membranes, and the respiratory system, drowsiness, fatigue, general malaise, and asthmatic symptoms (Becher et al. 1996; Kostiainen 1995).

Indoor exposure to these chemicals is considered widespread. The EPA has identified 300 VOCs in homes (Cooke 1991). In a study of VOCs in the indoor air of a number of households in Finland, clinical signs of VOCs disappeared after the elimination of a localized emission source (Kostiainen 1995).

Formaldehyde is a well-known VOC of great public concern (Samet et al. 1988). However, because of differences in measurement techniques, formaldehyde is not always included in studies of VOCs (Norback et al. 1995). This chemical was classified under the EPA's original weight-of-evidence rules as a probable (B1) human carcinogen based on limited evidence in humans and sufficient evidence in animals (EPA 1991b). The IRIS database also describes occupational studies showing significant associations between respiratory cancers and exposure to formaldehyde or formaldehyde-containing products, and nasal cancer in mice and rats exposed by inhalation to formaldehyde.

Cooke (1991) describes noncancer effects of formaldehyde in humans including irritation of the eyes and respiratory tract following acute-duration exposure. Cooke also concludes that acute exposures to high concentrations (37–125 mg/m^3) of formaldehyde can cause respiratory distress, inflammation of the lungs, pulmonary edema, and death.

3. Combustion Products

Combustion products represent a complex mixture of pollutants including carbon monoxide, carbon dioxide, nitrogen oxides, particulates, sulfur dioxide, and wood smoke. Carbon monoxide (CO) is a colorless, odorless gas that decreases the oxygen-carrying capacity of the blood (Cooke 1991). CO can cause neurological effects including headaches, dizziness, weakness, nausea, confusion, disorientation, and fatigue (EPA 1995b). At high concentrations death may occur. Carbon dioxide is a gas that can alter basic physiological functions at very high (> 30,000 ppm) concentrations (Cooke 1991).

Nitrogen oxides (NO, NO$_2$, and N$_2$O) are irritant gases (Cooke 1991). The acute effects of NO$_2$ on pulmonary function are well known (Cooke 1991). Acute effects include increased airway resistance in asthmatics and healthy individuals, and decreased pulmonary diffusing capacity. Chronic lung disease has been associated with long-term exposure to nitrogen dioxide. Samet et al. (1987) describe animal studies showing that NO$_2$ exerts adverse effects on lung defense mechanisms (i.e., mucociliary clearance and alveolar macrophage) and indicate that the effects have been demonstrated on the immune system.

4. Environmental Tobacco Smoke

Environmental tobacco smoke (ETS) is a complex mixture of gases and particles that has received considerable public attention in recent years (OSHA 1994). Components of both mainstream and sidestream smoke are quite numerous; primary components are respirable particulates, nicotine, polycyclic aromatic hydrocarbons, CO, acrolein, nitrogen dioxide, and many other chemicals (Samet et al. 1987). According to an OSHA assessment, the human health effects of ETS may include irritation of the eye and upper respiratory tract, pulmonary effects (e.g., lung function changes), cardiovascular effects (e.g., thrombus formation, vascular wall injury, aggravation of existing heart conditions, chronic heart disease), reproductive effects (e.g., low birth weight, miscarriage, increase in congenital abnormalities), and lung cancer (OSHA 1994).

C. Biological Contaminants

Biological contaminants represent a diverse array of biological agents that includes viruses, molds, mildew, house dust mites, fungal spores, algae, amoebae, arthropod fragments and droppings, and animal and human dander (Samet 1988; EPA 1995). Exposure to biological contaminants can cause numerous health effects such as allergic reactions (e.g., allergic rhinitis, asthma), infectious illnesses, hypersensitivity pneumonitis, humidifier fever, and Legionnaires' disease.

Hypersensitivity pneumonitis and humidifier fever are immunologically mediated diseases with lung symptomology (Samet et al. 1988). The acute form of hypersensitivity pneumonitis consists of fever, chills, cough, and dyspnea, while the chronic condition involves progressive dyspnea and lung function impairment. Fungi, bacteria, actinomycetes, amoebae, and nematodes have been identified as culprits. Legionnaires' disease is an acute bacterial infection resulting from indoor exposures to *Legionella pneumophila* (Samet et al. 1988). Rhinitis, coughing, sneezing, watery eyes, and asthma are some of the characteristic symptoms (EPA 1995b).

Approaches to the study of airborne contagious diseases, including outbreaks and epidemics, sampling during natural outbreaks, and experimental aerobiology have been discussed by Burge (1995). Burge notes that evidence that a disease is associated with indoor bioaerosols can be derived from: (1) case studies, or larger epidemiological studies, (2) sampling the air to demonstrate that airborne transport has occurred, or (3) experimental approaches (e.g., artificial transmission to animals or humans).

IV. LIMITATIONS OF THE APPLICATION OF HAZARD IDENTIFICATION TO INDOOR AIR POLLUTANTS

The identification of indoor air pollutants as potentially hazardous is complicated because of limitations and uncertainties inherent in the risk assessment process. The primary issues involve limitations of epidemiologic and animal studies, the nonspecificity of the symptomology of indoor air pollutants, and difficulties in the quantification of indoor air pollutant concentrations. These issues are discussed in more detail below.

A. Limitations of Epidemiologic Studies

As mentioned earlier, epidemiologic data, whenever available, are particularly useful in the hazard identification process. However, limitations that can affect epidemiologic studies include a small sample size, characteristics of a study population that are not representative of the population as a whole, the lack of statistical power, the presence of confounders, and uncertain exposure assessment. Studies that utilize questionnaires can be subject to selection and information bias. Misclassification errors regarding exposures and uncertain symptom registration can occur. Moreover, variables commonly examined in epidemiologic studies (e.g., subtle changes in lung function) are often prone to measurement error, and the relevancy of such changes may be difficult to interpret from a clinical perspective.

The size of the population under study is of particular importance in the identification of hazards associated with indoor air pollutants (Weiss 1993). Because of the relatively low levels of exposure to indoor air pollutants, and the limited variation in exposure to indoor air pollutants in members of the population, a very large number of subjects are required in a study to detect slight increases in the incidence of an adverse health effect. Other important considerations of epidemiological studies of indoor air pollutants include an accurate and unbiased assessment of a particular health outcome, and the selection of an unbiased sample of exposed and nonexposed individuals.

Confounding can be a serious problem in assessments of the associations between exposure to indoor air pollutants and health hazards. Temperature, humidity, barometric pressure, concomitant exposure to outdoor air pollutants, and cigarette smoking are some of the examples of confounders that are not usually controlled in studies of indoor air pollutants. The significance of confounders on the interpretation of epidemiologic data has been shown in a recent study by Moolgavkar and Luebeck (1996) on the association between particulate matter air pollution and mortality. For example, they show that the small risks associated with exposure to particulate matter could easily be attributed to residual confounding by copollutants. Moolgavkar and Luebeck (1996) also emphasize the impact of methodologic issues (e.g., modification of air pollution by seasonal effects), and the lack of appropriate statistical tools to assess the contribution of the particulate matter component. They concluded that it is not possible with the present evidence to show a convincing correlation between particulate air pollution and mortality.

Numerous indoor risk factors, such as age, gender, ethnicity, socioeconomic status, parental asthma, previous viral infection, hay fever, atopy, infant lung disease, low birth weight, geographic region of residence, and household water damage, have been identified as factors in asthma and wheezing. These symptoms are often linked with indoor air pollutants (Maier et al. 1997). The failure to adjust for risk factors can hinder interpretation of a study of indoor air pollution.

In recent years, there has been a growing awareness that psychological factors, such as differences in the perception of odor and discomfort, and psychological stress, may play an important role in the development of many of the nonspecific and vague symptoms often attributed to exposure to indoor air pollutants (Rothman and Weintraub 1995). There is evidence that stress, heavy work load, and conflicting demands can influence the number and severity of reported complaints encountered in the indoor environment (Nielsen et al. 1995). However, there are no well-designed, carefully controlled studies that have specifically established the extent of the impacts of such factors, or how to control for them in the design and performance of studies.

B. Nonspecificity of the Symptoms of Indoor Air Pollutants

One of the impediments encountered in the identification of the potential hazard of an indoor air pollutant is the nonspecific nature of the purported symptoms. The similarity of indoor air symptoms to common illnesses such as influenza, food poisoning, gastrointestinal disorders, Alzheimer's disease, angina, or brain deterioration can result in misdiagnosis of toxicity from indoor air pollution, and the underestimation of hazards from indoor air pollutants (Ammann 1987). Chronic-obstructive pulmonary disease may reflect a cumulative process in which air pollution is only one of the possible factors that can result in irreversible loss of lung function (Dockery 1993). As such, it is important to assess the potential health hazard of indoor air pollution with accuracy.

The application of hazard identification to understanding the phenomenon of indoor air "sensitivity," which often presents nonspecific, vague symptomology, is also complex. For instance, the problems of distinguishing between sensitivity resulting from indoor air exposure to chemicals and sensitivity resulting from exposure to bacteria, mites, foods, or allergens such as dust has been recognized (Henry et al. 1991). Moreover, other factors that may play a role in increased sensitivity in some individuals (e.g., multiple chemical sensitivity) such as comfort variables (i.e., heat and humidity), ventilation parameters, microbiological contaminations, and other airborne pollutants (e.g., CO, volatile organic chemicals, aldehydes, particles, pesticides) are largely ignored in indoor air studies (Pauluhn 1996). The problem is further compounded in that there are many types of chemical sensitivity (e.g., multiple chemical sensitivity and sick building syndrome) and the underlying mechanisms for these sensitivities remain elusive (Henry et al. 1991).

C. Difficulties in the Quantification of the Concentration of Indoor Air Pollutants

Although hazard identification requires the quantification of the concentration of a contaminant, the lack of precise measurements during the actual exposure period, or errors in the quantification of the concentration of a particular indoor air pollutant, can result in a failure to identify a hazard (Weiss 1993; Pauluhn 1996). The relevance of this issue is illustrated by the recent study of Pauluhn (1996) on the assessment of pyrethroids, a class of synthetic pesticides, following indoor use. It was found that measurement of deposited house dusts of the pesticide was a poor substitute for airborne dust measurements (Pauluhn 1996). Even under worst-case testing conditions (i.e., continuous brushing of the carpet for about nineteen hours in a bias-flow compartment) only a very small fraction of the pesticide-containing dust particles was found to be recovered airborne ($0.04\%/m^2$ per hr). Pauluhn (1996) concluded that state-of-the-art assessment of health hazards in the indoor environment based only on "vacuum cleaner" sampling is prone to a "high level of errors and misjudgment." It is noteworthy that vacuum cleaner sampling is often used in the study of indoor air pollutants and that such sampling may underestimate the potential hazards of indoor air pollutants.

Adequate quantification of the exposure to biological contaminants has been hampered by the lack of the development of standardized sampling methods, and problems related to the efficiency of collection by sampling apparatus (Samet et al. 1988). In addition, concentrations of biological agents can vary because of biological cycles and physical processes that influence the distribution of organisms in the air. Therefore, it is important to know in detail the specific species studied (including the life cycle), the collection efficiency of sampling apparatus, and the conditions under which sampling was conducted (Samet et al. 1988).

D. Limitations of Animal Studies

Studies of indoor air pollutants in experimental animals are also limited. A particular difficulty is in the choice of the most suitable animal species to study. Since animals may exhibit significant differences in the absorption and metabolism of a specific pollutant, cross-extrapolation of the identification of a hazard to humans may be inappropriate. Moreover, the results of animal toxicity studies are often difficult to use in predicting potential hazards in the most susceptible humans. The differences in nasal morphology and airflow dynamics among species should be considered for dosimetric adjustments. These limitations can result in problems of assigning weight of evidence to the potential hazard of an indoor air pollutant. The difficulties in obtaining indoor air samples of pollutants or appropriately simulating exposures also limits animal bioassay studies of the potential hazards of indoor air pollutants (Lewtas et al. 1993).

V. CRITICAL APPRAISAL OF THE DATA CONCERNING THE HEALTH HAZARDS OF INDOOR AIR POLLUTANTS

When the available epidemiologic data on indoor air pollution is examined as a whole, it is clear that many of the studies have failed to provide strong, definitive associations between exposure to indoor air pollutants and adverse health effects. In part, this reflects the lack of well-designed epidemiological studies that have controlled for numerous confounders and that have utilized appropriate statistical tools. Furthermore, there is a paucity of data regarding the long-term effects of exposure to low concentrations of indoor air pollutants, the relative roles of indoor vs. outdoor air pollutants, and the significance of the various constituents of complex indoor air pollutant mixtures in the manifestation of toxicological response. Finally, the mechanisms of toxicity of indoor air pollutants are not clear.

A review of the literature also reveals another problem, namely the consistency and validity of the available findings on indoor air pollution. Indeed, studies of the relation between exposure to indoor air volatile organic compounds (VOCs) and sick building syndrome have shown only a sparse or inconsistent association between observed VOC levels and health effects (Becher et al. 1996). Uncertain exposure assessment and symptom registration as well as limitations within study designs have been considered as contributing factors (Becher et al. 1996). As an example, it has been noted that the sets of VOCs selected for analysis in different studies are inconsistent, and the basis for the selections is unclear (Becher et al. 1996).

Likewise, no consistent evidence for a relationship between exposure to combustion products from gas stoves and excess respiratory symptoms and illnesses in children has been reported in any epidemiological studies (Samet et al. 1987). While a few studies have suggested effects of gas cooking on pulmonary function and respiratory symptoms, and acute respiratory illness in adults, potential confounders like cigarette smoking and chronic respiratory diseases were not considered in these studies (Samet et al. 1987). Inconsistent results have been reported in studies of the relation between indoor nitrogen dioxide exposure and respiratory health effects in children (Anto and Sunyer 1995).

Recent studies of ETS have also shown inconsistent relationships between passive smoking and wheezing and asthma (Samet et al. 1987). The inconsistency between workplace and spousal studies of ETS and lung cancer has been noted (LeVois and Layard 1994). In particular, they suggest that an estimate of ETS-lung cancer risk from female spousal smoking studies is inappropriate because of bias arising from spousal smoking study designs.

Studies regarding the assessment of hazards from biological contaminants are also limited. For instance, in studies of house dust mites it has been difficult to assess the relationship between the severity of asthma and exposure to dust mites, as well as determining the prevalence of house dust-mite-related asthma (Samet et al. 1988).

Although several studies suggest an increased frequency of respiratory symptoms among adults and children in damp houses (and consequently exposed to mold species), these studies have not considered the role of nonallergenic mechanisms (Becher et al. 1996). Such mechanisms include inhalation exposure to airborne toxic factors such as bacterial cell wall components and spores of toxin-producing molds with mycotoxins.

There is controversy regarding the health effects from exposure to particulate matter (Moolgavkar and Luebeck 1996). Issues of coherence, consistency, strength of association, linearity of exposure–response relationships, specificity, temporality, and biological plausibility have been raised. A lack of consistent association between symptom data and measures of particulate matter air pollution has been noted (Gamble and Lewis 1996). It has also been noted that individual-level study results of particulate matter are not coherent with time-series ecologic study results of hospital emissions (Gamble and Lewis 1996). These issues may also be pertinent to particulate matter in indoor environments. As mentioned earlier, particulate matter can exist in both outdoor and indoor environments, and many investigators realize the importance of measuring the relationships between outdoor and indoor environments. In addition, the potential mechanisms of possible causality between low levels of indoor air pollutants and toxicity have not been addressed. This includes potential interactive mechanisms among indoor air pollutants.

VI. SUMMARY

Various health hazards have been attributed to indoor air pollutants. Primary hazards of concern include cancer, irritative and respiratory effects, neurological effects, and developmental and reproductive toxicity. Several approaches are available to identify potential hazards associated with indoor air pollutants. These approaches constitute the hazard identification step of the risk assessment process.

Uncertainty is inherent in the hazard identification process, as a result of limitations of epidemiologic and animal studies, the nonspecificity of the symptomatology of indoor air pollutants, and the problems of inadequate quantification of the concentration of indoor air pollutants. These limitations and uncertainties are evident in much of the literature on indoor air pollutants. There is a need for more data concerning the potential hazards of indoor air pollutants following long-term exposure to low concentrations. More mechanistic data and a better understanding of the roles of various constituents of complex mixtures of indoor air pollutants would also be useful.

BIBLIOGRAPHY

American Conference of Governmental Industrial Hygienists. 1988. *Toxic Air Pollutant Guidelines: Review of Recent Progress and Problems*, American Conference of Governmental Industrial Hygienists: Cincinnati, OH.

Ammann, H.M. 1987. Effects of indoor air pollution on sensitive populations, *Clinical Ecology* V(1):15–21.

Anonymous. 1992. Indoor air pollution and acute respiratory infections in children, *Lancet* 339:396–398.

Anto, J.M., Sunyer, J. 1995. Nitrogen dioxide and allergic asthma: Starting to clarify an obscure association, *Lancet* 345/8947:402–403.

Auvinen, A., Makelaine, I., et al. 1996. Indoor radon exposure and risk of lung cancer: A nested case-control study in Finland, *Journal of the National Cancer Institute* 88(14):966–972.

Ayres, J.G. 1995. Epidemiology of the effects of air pollutants on allergic disease in adults, *Clinical and Experimental Allergy* 25(3):47–51.

Becher, R. et al. 1996. Environmental chemicals relevant for respiratory hypersensitivity: The indoor environment, *Toxicology Letters* 86:155–162.

Brown, R.C., Hoskins, J.A. 1992. Contamination of indoor air with mineral fibres, *Indoor Environment*, 1:61–68.

Burge, H.A. 1990. Risks associated with indoor infectious aerosols, *Toxicology and Industrial Health* 6(2):263–274.

Burge, H.A. 1995. *Bioaerosols*, Indoor Air Research Series, Center for Indoor Air Research, Lewis Publishers: Boca Raton, FL.

Burr, M.L., 1995. Pollution: Does it cause asthma?, *Journal of the British Paediatric Association* 72(5):377–379.

Carswell, F. 1995. Epidemiology of the effects of air pollutants on allergic disease in children, *Clinical and Experimental Allergy* 25(3):52–55.

Chapman, M.D. et al. 1995. Immunochemical and molecular methods for defining and measuring indoor allergens: In dust and air, *Pediatr. Allergy Immunol.* 6:8–12.

Cooke, T.F. 1991. Indoor air pollutants — A literature review, *Reviews on Environmental Health* 9:137.

Dockery, D. 1993. Epidemiologic study design for investigating respiratory health effects of complex air pollution mixtures, *Environmental Health Perspectives*, Supplement 101:187–191.

Dockery, D.W., Pope, C.A. 1994. Acute respiratory effects of particulate air pollution, *Annual Review of Public Health* 15:107–32.

Environmental Protection Agency (EPA). 1983. *Review and Evaluation of the Evidence for Cancer Associated with Air Pollution*, Report No. EPA-450/5-83-006, Office of Air Quality, Planning and Standards, U.S. Environmental Protection Agency, Research Triangle Park, NC.

Environmental Protection Agency (EPA). 1986a. Guidelines for carcinogen risk assessment, *51 FR 33992*.

Environmental Protection Agency (EPA). 1986b. Guidelines for mutagenicity risk assessment, *51 FR 34006*.

Environmental Protection Agency (EPA). 1986c. Guidelines for the health assessment of suspect developmental toxicants, *51 FR 34028*.

Environmental Protection Agency (EPA). 1991a. Guidelines for developmental toxicity risk assessment, *56 FR 63798*.

Environmental Protection Agency (EPA). 1991b. *Integrated Risk Information System (IRIS) Database, Appendix A, Reference dose (RfD): Description and Use in Health Risk Assessments*, Office of Health and Environmental Assessment, U.S. Environmental Protection Agency, Washington, DC.

Environmental Protection Agency (EPA). 1995a. Proposed guidelines for neurotoxicity risk assessment, *60 FR 52032*.

Environmental Protection Agency (EPA). 1995b. The Inside Story — A Guide to Indoor Air Quality, Report No. 402-K-93-007, U.S. Environmental Protection Agency, Washington, DC.

Environmental Protection Agency (EPA). 1996a. Proposed guidelines for carcinogen risk assessment, *61 FR 17959*.

Environmental Protection Agency (EPA). 1996b. Reproductive toxicity risk assessment guidelines, *61 FR 56273*.

Feron, V.J., Groten, J.P., et al. 1995. Toxicology of chemical mixtures: Challenges for today and the future, *Toxicology* 105:415–427.

Gamble, J.F., Lewis, R.J. 1996. Health and respirable particulate (PM_{10}) air pollution: A casual or statistical association?, *Environmental Health Perspectives* 104(8):838–850.

Gilmour, M.I. 1995. Interactions of air pollutants and pulmonary allergic responses in experimental animals, *Toxicology* 105:335–342.

Henry, C.J. et al. 1991. Approaches for assessing health risks from complex mixtures in indoor air: A panel overview, *Environmental Health Perspectives*, 95:135–143.

Husman, T. 1996. Health effects of indoor-air microorganisms, *Scandinavian Journal of Work Environment and Health* 22:5–13.

International Agency for Research on Cancer (IARC). 1993. *Effects of Indoor Air Pollution on Human Health*, Chapter 2, In: Volume 109:5–17, IARC Scientific Publications: Lyon, France.

Jarvis, D., Chinn, S., et al. 1996. Association of respiratory symptoms and lung function in young adults with use of domestic gas appliances, *Lancet* 347:426–431.

Kostiainen, R. 1995. Volatile organic compounds in the indoor air of normal and sick houses, *Atmospheric Environment* 29(6):693–702.

LeVois, M.E., Layard, M.W. 1994. Inconsistency between workplace and spousal studies of environmental tobacco smoke and lung cancer, *Regulatory Toxicology and Pharmacology* 19:309–316.

Lewtas, J. et al. 1993. Bioassay of Complex Mixtures of Indoor Air Pollutants, Volume 109:85–95, IARC Scientific Publications: Lyon, France.

Maier, W.C. et al. 1997. Indoor risk factors for asthma and wheezing among Seattle school children, *Environmental Health Perspectives*, 105:208–214.

Marbury, M.C., Maldonado, G., Waller, L. 1996. The indoor air and children's health study: Methods and incidence rates, *Epidemiology* 7(2):166–174.

Moolgavkar, S.H., Luebeck, E.G. 1996. A critical review of the evidence on particulate air pollution and mortality, *Epidemiology* 7:420–428.

National Research Council (NRC). 1983. *Risk Assessment in the Federal Government: Managing the Process*, Prepared by the Committee on Institutional Means for Assessment of Risk to Public Health, Commission on Life Sciences, National Academy Press: Washington, DC.

Newman-Taylor, A. 1995. Environmental determinants of asthma, *Lancet* 345:296–299.

Nielsen, G.D., Alarie, Y., et al. 1995. Possible mechanisms for the respiratory tract effects of noncarcinogenic indoor-climate pollutants and bases for their risk assessment, *Scandinavia Journal of Work Environment and Health* 21:165–178.

Norback, D., Bjornsson, E., et al. 1995. Asthmatic symptoms and volatile organic compounds, formaldehyde, and carbon dioxide in dwellings, *Occupational and Environmental Medicine* 52:388–395.

Occupational Safety and Health Administration (OSHA). 1994. *29 CFR Parts 1910, 1915, 1926, 1928.*

Office of Science and Technology Policy. 1984. Chemical carcinogens: A review of the science and its associated principles, *49 FR 21594-21661.*

Office of Science and Technology Policy. 1985. Chemical carcinogens: A review of the science and its associated principles, *50 FR 10372-10442.*

Otto, D.A., Hudnell, H.K. 1993. The use of visual and chemosensory evoked potentials in environmental and occupational health, *Environmental Research* 62:159–171.

Pauluhn, J. 1996. Risk assessment of pyrethroids following indoor use, *Toxicology Letters* 88:339–348.

Pope, C.A. et al. 1995. Review of epidemiological evidence of health effects of particulate air pollution, *Inhalation Toxicology* 7:1–18.

Robison, T.W., Zhou, H., et al. 1996. Generation of glycoaldehyde from guinea pig airway epithelial monolayers exposed to nitrogen dioxide and its effects on sodium pump activity, *Environmental Health Perspectives* 104(8):852–856.

Rothman, A.L., Weintraub, M.I. 1995. The sick building syndrome and mass hysteria, *Neurologic Clinics* 13(2):405–412.

Samet, J.M. et al. 1987. Health effects and sources of indoor air pollution, Part I, *American Review of Respiratory Disease* 136:1486–1508.

Samet, J.M. et al. 1988. Health effects and sources of indoor air pollution, Part II, *American Review of Respiratory Disease* 137:221–242.

Samet, J.M., Speizer, F.E. 1993. Assessment of health effects in epidemiologic studies of air pollution, *Environmental Health Perspectives* 101:149–154.

Samet, J.M., Utell, M.J. 1990. The risk of nitrogen dioxide: What have we learned from epidemiology and clinical studies?, *Toxicology and Industrial Health* 6(2):247 (abstract).

Sidhu, K.S., Hesse, J.L., et al. 1993. Indoor air: Potential health risks related to residential wood smoke, as determined under the assumptions of the US EPA risk assessment model, *Indoor Environment* 1993(2):92–97.

Simmons, J.E. 1995. Chemical mixtures: Challenge for toxicology and risk assessment, *Toxicology* (abstr.) 105(1995):111–119.

Solomon, W.R. 1990. Airborne microbial allergens: Impact and risk assessment, *Toxicology and Industrial Health* (abstr.) 6(2):309.

Sram, R.J., Benes, I., et al. 1996. Teplice program—the impact of air pollution on human health, *Environmental Health Perspectives* 104(4):699–714.

Stolwijk, J.A.J. 1987. Multipollutant Indoor Exposures and Health Responses: Epidemiological Approaches, for presentation at the 80th annual meeting of APCA, New York, NY, June 21–26, 1987.

Tepper, J.S., Moser, V.C., et al. 1995. Toxicological and chemical evaluation of emissions from carpet samples, *American Industrial Hygiene Association Journal* 56:158–170.

Vogt, R.F. 1991. Use of laboratory tests for immune biomarkers in environmental health studies concerned with exposure to indoor air pollutants, *Environmental Health Perspectives* 95:85–91.

Volkmer, R.E., Ruffin, R.E., et al. 1995a. The prevalence of respiratory symptoms in South Australian preschool children. I. Geographic location, *Journal of Paediatric Child Health* 31:112–115.

Volkmer, R.E., Ruffin, R.E., et al. 1995b. The prevalence of respiratory symptoms in South Australian preschool children. II. Factors associated with indoor air quality, *Journal of Paediatric Child Health* 31:116–120.

Weiss, N.S. 1993. Complex mixtures and indoor air pollution: Overview of epidemiologic methods, *Environmental Health Perspectives* 101:179–181.

Wheeler, C. 1990. Exposure to man-made mineral fibers: A summary of current animal data, *Toxicology and Industrial Health* (abstr.) 6(2):293.

Wolff, S.K., Hawkins, N.C., et al. 1990. Selecting experimental data for use in quantitative risk assessment: An expert judgment approach, *Toxicology and Industrial Health* (abstr.) 6(2):275.

Zummo, S.M., Karol, M.H. 1996. Indoor air pollution: Acute adverse health effects and host susceptibility, *Environmental Health* 58(6):25–29.

CHAPTER **4**

Dose–Response Assessment — Quantitative Methods for the Investigation of Dose–Response Relationships

Suresh H. Moolgavkar

CONTENTS

I. INTRODUCTION

The estimation of exposure–response or dose–response relationships is a prerequisite for a rational approach to the setting of standards for human exposures to

1-56670-323-9/99/$0.00+$.50
© 1999 by CRC Press LLC

potentially toxic substances. In many instances when human epidemiologic data are not available, standards are based on assessment of toxic responses in experimental data followed by extrapolation of risks to humans. Additionally, experiments in animals are often carried out at high exposure levels so that the experiments have the requisite statistical power. The resultant issues of interspecies and low-dose extrapolation are among the most contentious scientific issues of the day.

In the past few decades, a vast biostatistical literature has appeared on exposure–response and dose–response analyses. Summarizing this literature in a single chapter is a formidable task. Although many of the same methods can be used for experimental and epidemiologic studies, this chapter focuses on statistical methods that have been developed for analyses of epidemiologic studies, and on biologically based mathematical models for analyses of data in which the end point of interest is cancer. Physiologically based pharmacokinetic (PBPK) models, developed to investigate the relationship of exposure to dose by consideration of the uptake, distribution, and disposal of agents of interest, will be discussed only briefly. This is not because these models are considered to be unimportant, but because this subject is outside the author's area of expertise. Interspecies differences in response to exposure to environmental agents can often be explained, at least partially, in terms of differences in uptake and distribution of the agent. Thus, PBPK models have advanced broadly our understanding of differential species toxicology and can be considered important tools in risk assessment.

For risk assessment, epidemiologic studies offer two obvious advantages over experimental studies. Firstly, since the studies are done in the species of ultimate interest, the human, the difficult problem of interspecies extrapolation is finessed. Secondly, most epidemiologic studies are done at levels of exposure that are much closer to typical exposures in free-living human populations than is possible with experimental studies. It is true that often epidemiologic studies are conducted in industrial cohorts, which are typically exposed to higher levels of the agent of interest than the general population. Nonetheless, the levels of exposure, even in industrial cohorts, are much closer to those in the general population than the exposures used in experimental studies. Some of what epidemiologic studies gain in the way of relevance over experimental studies is given up in precision, however. It is generally true that both exposures and disease outcomes are measured with less precision in epidemiologic studies than in laboratory studies, leading, possibly, to bias in the estimate of risk and the shape of the dose–response curve. Exposure measurement error is now widely regarded as being an important issue in analyses of epidemiologic data. Another potential problem for risk assessment arises from the fact that human populations, especially industrial cohorts, are rarely exposed to single agents. When exposure to multiple agents is involved, the effect of the single agent of interest is often difficult to investigate. This fact is of particular relevance for air pollution because it is generally a complex mixture of toxic agents. The role of any single component of the mixture can be difficult to study.

Epidemiologic studies that can be used to investigate dose–response relationships are classified into three broad categories. The cohort study is, at least conceptually, close to the traditional experimental study in that groups of exposed and unexposed individuals are followed in time and the occurrence of disease in the two groups compared. In the case-control study, relative risks are estimated from cases of the disease under investigation and suitably chosen controls. In these two types of study, information on exposures and disease is available on an individual basis for all subjects enrolled in the study. In a third type of study, the ecological study, information is available only on a group basis. Ecological studies have generally been looked upon with disfavor by epidemiologists for reasons that have been extensively discussed elsewhere (Greenland and Morgenstern 1989; Greenland and Robins 1994). Nonetheless, they can provide useful information and, particularly, in air pollution epidemiology, they have played a central role in recent times. Another type of study, in which disease outcome and some confounders are known on an individual basis and others together with exposure to air pollutants are known only on a group basis, has recently played an important role in air pollution epidemiology. There is currently no generally accepted term for such studies, which share attributes of the cohort study and the ecological study. Such studies are called here hybrid studies. Because epidemiologic studies are observational (i.e., groups of subjects cannot randomly be assigned to one exposure group or another), careful attention must be paid to controlling factors that may bias estimates of risk. Thus, controlling for what epidemiologists call "confounding" is of paramount importance both in the design and analyses of epidemiologic studies.

Within the last two decades sophisticated statistical tools have been developed for the analyses of epidemiologic data. Many of these methods fall under the rubric of the so-called relative risk regression models. Additionally, recent research in air pollution epidemiology has exploited regression methods for analyses of time-series of counts. Both parametric and semiparametric Poisson regression models have been developed for analyses of these data. Special methods are required when multiple observations are made on the same individual, as is done in panel studies, or in the same geographic location, as is done with Poisson regression analyses of time-series of counts. Account must then be taken of serial correlations in the observations. Various statistical methods are used to address this issue. Finally, when the health effect of interest is cancer, stochastic models based on biological considerations can be used for data analyses. These models provide a useful complement to the more empirical statistical approaches to data analyses. Each of these approaches will be discussed briefly in this chapter.

II. MEASURES OF DISEASE FREQUENCY AND MEASURES OF EFFECT

When discussing dose– or exposure–response relationships it is important to define clearly what response one is talking about. Often the term *dose–* or *exposure–response* is used with no indication of what *response* means. In order to define response precisely it is important to have a clear idea of the various commonly used measures of disease frequency and of effect. Perhaps the most fundamental measure of disease frequency is the incidence rate, also called the hazard rate in the statistical literature. The incidence or hazard rate measures the rate (per person per unit time) at which new cases of a disease appear in the population under study. Because the incidence rates of many chronic diseases, including cancer, vary strongly with age, a commonly used measure of frequency is the age-specific incidence rate, usually reported in five-year age categories. For example, the age-specific incidence rate per year in the five-year age group 35–39 may be estimated as the ratio of the number of new cases of cancer occurring in that age group in a single year to the number of individuals in that age group who are cancer free at the beginning of the year. Strictly speaking, the denominator should be not the total number of individuals who are cancer free at the beginning of the year but the person-years at risk during the year. This is because some individuals contribute less than a full year of experience to the denominator, either because they enter the relevant population after the year has begun (for example, an individual may reach age 35 sometime during the year) or because they may leave the population before the year is over (for example, an individual may reach age 40, die, or migrate during the year). Mathematically, the concept of incidence rate is an instantaneous concept, and is most precisely defined in terms of the differential calculus. A precise definition of the concept is given in the next section, and the reader is referred to texts on survival analysis (e.g., Kalbfleisch and Prentice 1980; Cox and Oakes 1984) for further details.

Another commonly used measure of disease frequency is the probability that an individual will develop disease in a specified period of time. For risk assessment, interest is most often focused on the lifetime probability, often called lifetime risk of developing disease. Here, *lifetime* is arbitrarily defined in the U.S. usually as 70 years. The incidence (or hazard) rate and probability of developing disease are related by a simple formula. This relationship is expressed by the following equation:

$$P(t) = 1 - \exp\left(-\int_0^t I(s)\,ds\right)$$

where P(t) is the probability of developing the disease of interest by age t, and I(s) is the incidence or hazard rate at age s. Note that although the probability of disease, P(t), is called cumulative incidence in some epidemiology textbooks (Rothman 1986), the integral

$$\int_0^t I(s)\,ds$$

is actually the cumulative incidence. When the incidence rate is small, as is true for most chronic diseases, the probability of disease by time t, P(t), is approximately equal to the cumulative incidence,

$$P(t) \approx \int_0^t I(s)\,ds$$

The impact of an environmental agent on the risk of disease can be measured on either the absolute or the relative scale. The last two decades have seen an explosion of statistical literature on relative measures of risk, which can be estimated in both case-control and cohort studies. Let I_e be the incidence rate in the exposed population and I_u be the incidence rate in the unexposed population. Then the relative incidence (relative risk) is defined by

$$RR = I_e/I_u$$

A closely related measure is excess relative risk, which is defined as

$$ERR = (I_e - I_u)/I_u = RR - 1$$

Yet another measure of risk is the attributable or etiologic fraction, which is defined as

$$AF = (I_e - I_u)/I_e = (RR - 1)/RR$$

The AF is the fraction of incident cases in the exposed population that would not have occurred in the absence of exposure, and "can be interpreted as the proportion of exposed cases for whom the disease is attributable to the exposure" (Rothman 1986). In most regression analyses of epidemiologic data, RR is modeled either as a "multiplicative" or an "additive" function of the covariates of interest. Since RR is readily estimated from both case-control and cohort studies, the various measures of effect discussed above which are functions of RR alone can be estimated.

On the absolute scale, the impact of an agent can be measured simply by the difference of incidence rates (or probabilities) among exposed and nonexposed subjects. Absolute measures of risk cannot be estimated from case-control studies without ancillary information (Rothman and Greenland 1998).

The impact of an environmental agent on the risk of disease on a population will depend not only on the strength of its effect in the exposed subpopulation, but also on how large this subpopulation is. Even if the agent is a very potent carcinogen, its impact on the cancer burden of the entire population will be small if only a small fraction of the population is exposed. On the other hand, if exposure to a weak carcinogen is widespread, the population impact could be substantial. A measure of risk that attempts to quantify the population burden of disease due to a specific exposure is the population attributable fraction, PAF, which is defined as the fraction of all cases in the population that can be attributed to the exposure, and is given by the expression

$$PAF = (I_T - I_u)/I_T$$

where I_T is the incidence in the total population. In addition to the RR, estimation of the PAF requires information on the fraction of the population exposed to the agent of interest (see Rothman 1986). The PAF can be estimated directly from case-control data only if the controls are a random sample from the population (Rothman and Greenland 1998). When the RR associated with exposure to an agent is high and the exposure is widespread, a major fraction of disease in the population can be attributed to the agent. For example, it has been estimated that approximately 84% of all lung cancers and 43% of all bladder cancers in Australian men in 1992 could be attributed to cigarette smoking (English et al. 1995).

The calculation of the PAF can be extended to situations where there are multiple levels of exposure (by considering each level in turn and adding up the PAFs) or where the exposure is a continuous variable rather than a categorical one (by creating discrete categories such as quartiles or quintiles of exposure, or by using regression models). Joint effects of several exposures may be considered similarly. In the case of two or more exposures, the separate PAFs may be calculated for each exposure while ignoring the other exposures, or a combined PAF may be calculated by considering all possible combinations of exposures, calculating the PAF for each and adding up. When two or more exposures are involved, the sum of the separate PAFs will frequently exceed the combined PAF calculated in this way and may actually exceed 100%. The reason for this is clear: cases that occur in the joint exposure categories are counted multiple times when PAFs for single exposures are computed, once for each exposure in the joint exposure category. Attribution of causation in the case of joint exposures is best done by considering all possible combinations of exposures. For example, with two exposures, attribution of causation may be summarized by subdividing the cases into those that can be considered as being caused by the combination of the two agents, each agent exclusively, or neither agent (Enterline 1983). For a more advanced treatment of PAFs, see Bruzzi et al. (1985), Wahrendorf (1987), Benichou (1991), and Greenland and Drescher (1993).

III. CONFOUNDING

A detailed discussion of confounding, a concept of central importance in epidemiology, is outside the scope of this chapter. Confounding arises in epidemiologic studies as a consequence of the fact that these are observational (not randomized). Suppose one is interested in alcohol as a possible cause of oral cancer. Suppose that an epidemiologic study shows an association between alcohol consumption and oral cancer. That is, suppose the incidence of oral cancer in the subpopulation of individuals that imbibes alcohol is higher than the incidence of oral cancer in the subpopulation of teetotalers. The crucial question then is the following: Could the association between alcohol consumption and oral cancer be "spurious" in the sense that it is due to another agent that is itself a cause of oral cancer, and more likely to be found in the subpopulation of alcohol imbibers than in the subpopulation of teetotalers? One example of such an agent is tobacco smoke. Individuals who imbibe alcohol are more likely than teetotalers to be smokers. Moreover, smoking is strong risk factor for oral cancer. Thus the observed association between alcohol consumption and oral cancer may actually be due to the association between smoking and alcohol consumption. In a study of oral cancer and alcohol, tobacco smoke is a confounder.

As shown by this example, confounding is the distortion of the effect of the agent of interest by an extraneous factor. To be a confounder, a factor must satisfy two conditions. First, the putative confounder must be a risk factor for the disease in the absence of the agent of interest. Second, the putative confounder must be associated with the exposure of interest in the population in which the study is conducted. Sometimes a third condition (Rothman and Greenland 1998) is added—the putative confounder must not be an intermediate step in the pathway between exposure and disease. While these three criteria define a confounder for most epidemiologists, other definitions which are close but not identical to the definition given here have been given by biostatisticians. These are usually couched in terms of collapsibility of contingency tables. For a more detailed discussion, the reader is directed to Greenland and Robins (1986) and Rothman and Greenland (1998).

Confounding in epidemiologic studies can be addressed in one of two ways—it can be prevented by appropriate study design or controlled by appropriate analyses. The specific methods used depend upon the type of epidemiologic study. The reader is referred to recent texts (Rothman and Greenland 1998) for details.

The main statistical tools for exposure– and dose–response analyses of epidemiologic data will now be discussed briefly. Many of these methods can be used for analyses of experimental data as well.

IV. EMPIRICAL STATISTICAL METHODS

Because most epidemiologic studies are observational, issues of sampling and data analysis are particularly important to assure appropriate interpretation of results in the presence of possible confounding. Some of the main statistical tools developed over the last few decades to address these issues are discussed below.

A. Relative Risk Regression Models

The development here will follow that in the paper by Prentice et al. (1986). Although this paper was written over a decade ago, it lays out the basic framework for these models. The concept of hazard function was introduced above as being the appropriate statistical concept that captures the epidemiologic idea of an incidence rate. A more precise definition of this concept follows. Consider a large, conceptually infinite, population that is being followed forward in time, and about which one wishes to draw inferences regarding the occurrence of some health related event, generically referred to as a "failure." Typically one is interested in relating the failure to preceding levels of one or more risk factors, such as genetic and lifestyle factors and exposure to external agents, collectively referred to as covariates. Let z(t) denote the vector of covariates for an individual at time t. Time may be the age of the individual, or, in some settings, it may be more natural to consider other specifications, such as time from a certain calendar date, or duration of employment in a specific occupation. Let T denote the time of failure for a subject, and suppose that

$Z(t)$ represents the covariate history up to time t. Then the population frequency of failure, which may be thought of as the probability of failure, in a time interval t to $t + \Delta$ with covariate history $Z(t)$, will be denoted by $P[t + \Delta | Z(t)]$. The hazard or incidence function (which, if failure refers to death, is often called the force of mortality) is then defined by

$$h[t; Z(t)] = \lim_{\Delta \to 0} P[t + \Delta \,|\, Z(t); T \geq t] \Delta^{-1} = P'[t \,|\, Z(t)] \big/ \big(1 - P[t \,|\, Z(t)]\big)$$

In order to simplify notation, the dependence of h, P, etc. on the covariate history $Z(t)$ will be suppressed unless this is not clear from the context. Thus, for example, $h[t; Z(t)]$ will be written as $h(t)$. An intuitive interpretation of the hazard is that it is the rate of failure at time t among those who have not failed up to that time.

Now suppose that one is interested in the incidence of failures among individuals with a specific covariate history, $Z(t)$. For example, one may be interested in the incidence among individuals who are exposed to certain environmental agents thought to be associated with the disease under investigation. Let $Z_0(t)$ represent some standard covariate history; for example, $Z_0(t)$ could be thought of as the covariate history among those not exposed to the agents of interest. One can then write

$$h[t; Z(t)] = h_0(t) \, RR[t; Z(t)]$$

where $h_0(t) = h[t; Z_0(t)]$ and $RR[t; Z(t)]$ denotes the relative risk of failure at time t associated with covariate history $Z(t)$.

Relative risk regression models attempt to describe risks in populations by focusing on the relative risk function. Various functional forms for RR have been used, the most commonly used being "multiplicative" and "additive" functions of the covariates. The multiplicative model is given by

$$RR[t; Z(t)] = \exp(\beta_1 z_1 + \beta_2 z_2 + \ldots + \beta_n z_n)$$

and the additive model by

$$RR[t; Z(t)] = 1 + \beta_1 z_1 + \beta_2 z_2 + \ldots + \beta_n z_n$$

where z_1 through z_n are the covariates of interest and the β's are parameters to be estimated from the data. Note that the additive model posits that the relative risk is a linear function of the exposures of interest and that the effect of joint exposures is additive. The multiplicative model posits that the logarithm of relative risk is a linear function of the exposures and that the effect of joint exposures is multiplicative. Quite often the relative risk cannot be adequately described by either a multiplicative or an additive model. For example, the relative risk associated with joint exposure to radon and cigarette smoke is greater than additive but less than multiplicative (BEIR IV 1988). Various mixture models have been proposed (Thomas 1981; Breslow and Storer 1985; Guerrero and Johnson 1982) to address such situations.

The use of these models presents special statistical problems (Moolgavkar and Venzon 1987; Venzon and Moolgavkar 1988).

There is a vast biostatistical literature on the application of relative risk regression models to the analyses of various study designs encountered in epidemiology. It is outside the scope of this chapter to review this literature. The interested reader is referred to the appropriate publications (Breslow and Day 1980; Breslow and Day 1987).

In the field of air pollution epidemiology, relative risk regression models were used for analyses of two important studies of the long-term effects of air pollution on health. These are the Harvard Six Cities Study (Dockery et al. 1993) and the ACS II study (Pope et al. 1995). In these studies, cohorts of individuals were assembled from cities with different pollution profiles and information collected on certain life-style factors, such as cigarette smoking. These individuals were then followed and their mortality experience recorded. The authors of these studies refer to them as cohort studies. There is, however, an important element of the ecologic design to these studies. The exposure of interest, namely air pollution, is measured not on the individual level, but on the level of the city. That is, because information on concentrations of pollutants is available only from central monitoring stations, exposure to air pollution is assumed to be identical for all study subjects in a city. Because these studies combine elements of the cohort design with ecologic design, the term *hybrid studies* has been coined by this author for designs of this type. This study design can pose formidable problems in the interpretation of the results of analyses (Moolgavkar and Luebeck 1996).

B. Poisson Regression

Quite often information is available, not on individual members of a study cohort, but on subgroups that are reasonably homogeneous with respect to important characteristics, including exposure, that determine disease incidence. As a concrete example, consider the well-known British doctors' study of tobacco smoking and lung cancer. For the cohort of individuals in this study, information on the number of lung cancer deaths is cross-tabulated by daily level of smoking (reported in fairly narrow ranges) and five-year age categories. Another well-known example is provided by the incidence and mortality data among the cohort of atomic bomb survivors, for which the numbers of cancer cases are reported in cross-tabulated form by (ranges of) age at exposure, total dose received (in narrow ranges) and by five-year attained age categories. When data are presented in this way the method of Poisson regression is often used for analyses. Only a very brief outline of the method is given here. For more details the reader is referred to the standard text by McCullagh and Nelder (1989).

For Poisson regression, the number of events of the outcome of interest (death or number of cases of disease) in each cell in the cross-tabulated data is assumed to be distributed as a Poisson random variable with expectation (mean) that is a function of the covariates of interest. The numbers of events in distinct cells of the cross-tabulated data are assumed to be independent. Suppose that the data are presented in I distinct cross-tabulated cells, and let E_i be the expectation of the number of events in cell I. Suppose that the observed number of events in cell I is O_i. Then under the assumption that the number of events is Poisson distributed, the likelihood of the data is

$$L = \Pi_i \{E_i^{O_i} \exp(-E_i)\}/O_i!$$

where the product is taken over all the cells in the cross-tabulated data. The expectations E_i are made functions of the covariates of interest. Generally, $\log E_i$ is modeled as a linear function of the covariates. More elaborate functions have been used, however, for example in the analyses of the atomic bomb survivors data (BEIR V 1990) and the analyses of lung cancer in cohorts of underground miners (BEIR IV 1988; DHHS 1994). The expectation has been modeled as well by the hazard function of biologically based carcinogenesis models (Moolgavkar et al. 1989). Whatever the model form for the expectation, the parameters are estimated by maximizing the likelihood function.

Poisson regression models have played a prominent role in recent analyses of associations between indices of air quality in various urban areas and health outcomes such as mortality (Schwartz and Dockery 1992; Schwartz 1993) and hospital admissions (Burnett et al. 1994; Moolgavkar et al. 1997) for specific causes (respiratory disease, heart disease). These studies purport to investigate the acute effects of air pollution in contrast to the hybrid studies referred to above, which investigated the long-term effects of air pollution. In these analyses, daily counts of events (deaths or hospital admissions) in a defined geographical area are regressed against levels of air pollution as measured at monitoring stations in that area. Explicitly, the number of events on any given day is assumed to be a Poisson random variable, the expectation of which depends upon indices of air quality and weather on the same or previous days. In this type of study, inferences regarding the association of air pollution with the health events of interest depend upon relating fluctuations in daily counts of events to levels of air pollution on the same or previous days. As indicated above, in the simplest form of Poisson regression the logarithm of the expectation is a linear function of the covariates. This restriction on the shape of the exposure–response function may not be appropriate, and recently more flexible methods that make no assumptions regarding the shape of this relationship have been introduced for analyses of these data (Health Effects Institute 1995). An important difference between Poisson regression analyses of air pollution data and the other examples given above (e.g., analysis of the atomic bomb survivor data) is that in the air pollution data information on exposure is available only from central monitors of air quality. It is not possible to form strata of individuals with like exposures within a narrow range. It is not possible, therefore, to investigate the number of deaths or hospital admissions among individuals similarly exposed. This fact makes this type of study of air pollution an ecological study in that exposures and outcomes are known only on the group level, and it is not clear that the number of events is related to the level of exposure.

V. BIOLOGICALLY BASED MODELS

Biologically based models for the process of carcinogenesis have been in use for analyses of epidemiologic and experimental data for the past four decades. When the response of interest is cancer, these models provide a useful complement to the empirical statistical methods briefly described above for the analyses of data. Because the parameters of the model have direct interpretation in biological terms, analyses using these models may lead to testable hypotheses. These models also provide a framework within which the process of carcinogenesis can be viewed, and help break up a complex problem into simpler component pieces. The models are particularly useful for analyses of data with complicated patterns of exposure to environmental agents, as typically occurs with exposures in occupational cohorts where workers may switch jobs often. The linearized multistage procedure, which has been used as a default by the United States Environmental Protection Agency (EPA), is based on an early stochastic model, the Armitage-Doll multistage model.

VI. MULTISTAGE MODELS

Current understanding of carcinogenesis as a complex multistage process is based on observations from histopathological, epidemiologic, and molecular biological studies. Disruption of normal cell proliferation is the sine qua non of the malignant state. Conversely, there is accumulating evidence that the kinetics of cell division, cell differentiation (or death), and apoptosis of normal and premalignant cells are important in the carcinogenic process (Cohen and Ellwein 1990). Increases in cell division rates may lead to increases in the rates of critical mutational events, and an increase in cell division without a compensatory increase in differentiation or apoptosis leads to an increase in the size of critical target cell populations. These observations indicate that carcinogenesis involves successive genomic changes, some of which may result in disruption of normal cellular kinetics and facilitate the acquisition of further mutations. The number of necessary genomic changes required for malignant transformation is not known with certainty for any tumor, although it is thought to be at least two.

The following fundamental assumptions underlie the models considered here:

(1) cancers are clonal (i.e., malignant tumors arise from a single malignant progenitor cell);
(2) each susceptible (stem) cell in a tissue is as likely to become malignant as any other;
(3) the process of malignant transformation in a cell is independent of that in any other cell; and
(4) once a malignant cell is generated, it gives rise to a detectable tumor with probability 1 after a constant lag time.

The last two assumptions are clearly false, and are made for mathematical convenience. Methods for relaxing these assumptions are currently being investigated

(Yang and Chen 1991; Luebeck and Moolgavkar 1994). A mathematical review of some carcinogenesis models can be found in the book by Tan (1991).

A. The Armitage–Doll Multistage Model

The Armitage–Doll model, which has been used extensively in the last four decades, was first proposed to explain the observation that, in many human carcinomas, the age-specific incidence rates increase roughly with a power of age. The Armitage–Doll model postulates that a malignant tumor arises in a tissue when a single susceptible cell in that tissue undergoes malignant transformation via a finite sequence of intermediate stages, the waiting time between any stage and the subsequent one being exponentially distributed. Schematically, the model may be represented as follows:

$$E_0 \rightarrow E_1 \rightarrow ...E_{n-1} \rightarrow E_n$$

Here E_0 represents the normal cell, and E_n the malignant cell. Suppose that a cell moves from stage E_j to stage E_{j+1} with transition rate λ_j. Precisely, this means that the waiting time distribution for a cell to move from stage E_j to stage E_{j+1} is exponential with parameter λ_j. Let $p_j(t)$ represent the probability that a given cell is in stage E_j by time t. Then, $p_n = p(t)$ is the probability that the cell is malignantly transformed by time t, and the expression for the hazard, h(t), is given by $h(t) = N p'(t)/(1-p(t))$, where N is the number of susceptible cells in the tissue. In the usual treatment of the multistage model, two approximations are usually made at this point. First, at the level of the single cell, malignancy is a very rare phenomenon. Thus, for any cell, p(t) is very close to zero during the life span of an individual, and h(t) is approximately equal to $Np'(t)$. An explicit expression for $Np'(t)$ in terms of the transition rates λ_j is given in Moolgavkar (1978, 1991). Expanding $p'(t)$ in a Taylor series, one obtains

$$h(t) \approx Np'(t) = N\lambda_0\lambda_1 ... \lambda_{n-1}t^{n-1}\{1 - \text{mean}(\lambda)t + f(\lambda, t)\}/(n - 1)!$$

where mean(λ) is the mean of the transition rates and $f(\lambda, t)$ involves second and higher order moments of the transition rates. Retention of only the first nonzero term (this is the second approximation) in this series expansion leads to the Armitage–Doll expression, namely

$$h(t) \approx N\lambda_0\lambda_1 ... \lambda_{n-1}t^{n-1}/(n-1)!$$

Thus, with the two approximations made, this model predicts an age-specific incidence curve that increases with a power of age that is one less than the number of distinct stages involved in malignant transformation.

It is immediately obvious from the model that, given sufficient time, any susceptible cell eventually becomes malignant. Further, since the waiting time distribution to malignant transformation is the sum of n exponential waiting time distributions, it follows that h(t) is a monotone increasing function. Moreover, one can show that h(t) has a finite asymptote:

$$\lim_{t \to \infty} h(t) = N\lambda_{min}$$

where λ_{min} is the minimum of the transition rates. By contrast, the Armitage–Doll approximation increases without bound, and thus becomes progressively worse with increasing age (Moolgavkar et al., in press).

In order to model the action of environmental carcinogens, one or more of the transition rates can be made functions of the dose of the agent in question, where dose is to be thought of as the effective dose at the target tissue. Usually, the transition rates are modeled as linear functions of the dose, so that $\lambda_j = a_j + b_j d$ for one or more j. The assumption of first order kinetics may be justified, at least for carcinogens that interact directly with DNA to produce mutations. Then, using the Armitage–Doll approximation, the hazard function at age t and dose d can be written as

$$h(t,d) = g(d)t^{n-1}$$

where $g(d) = \{N \Pi_j(a_j + b_j d)\}/(n-1)!$ and the probability of tumor is approximately given by

$$P(t,d) = 1 - \exp[-g(d)t^n/n]$$

Note that g(d) is a product of linear terms. It is in this form, called the linearized multistage model, that the Armitage–Doll model is applied to the problem of low-dose extrapolation. Generally, the proportion of animals developing tumors at a specified fixed age at each of three different dose levels is known. The linearized multistage model is fitted to the data and the estimated parameters used to extrapolate risk to lower doses. There are formally at least two problems with this procedure. First, as noted above, the Armitage–Doll approximation may be poor when the probability of tumor is high as is the case in the usual animal experiments used for risk assessment. Second, in statistical fitting of the linearized multistage model, g(d) is treated as a general polynomial whereas it is really a product of linear terms.

B. The Two-Mutation Clonal Expansion Model

This model, referred to as the two-mutation model, is best interpreted within the framework of the initiation-promotion paradigm in chemical carcinogenesis. One of the features of this model is that, unlike the Armitage–Doll model, it takes explicit account of cell division and death of intermediate cells on the pathway to malignancy. The postulates of the model are as follows. Each tissue contains a pool of cells (believed to be stem cells) susceptible to malignant transformation. Each of these cells has a nonzero probability of sustaining a (series of) critical heritable change(s) to its genome leading to partial abrogation of cellular control of cell division and apoptosis (programmed cell death). A cell that has sustained such an event is called an initiated cell. Initiation is a rare event, the probability of which may be enhanced or decreased by external agents. An agent that enhances this probability is called an initiator. Most DNA damaging agents, such as ionizing radiation, are initiators. Initiated cells may divide or die and may give rise to populations of initiated cells. A primary initiated cell together with its daughter cells is called a clone of initiated cells. The process of clonal growth of initiated cells is called promotion. Such clonal outgrowths of initiated cells give rise to premalignant lesions. Examples of such lesions are papillomas of the mouse skin, enzyme altered foci in the rodent liver, and adenomatous polyps in the human colon. Rates of cell division and apoptosis of initiated cells may be enhanced or decreased by external agents. Any agent that facilitates the clonal growth of initiated cells is called a promoter. Many hormones are endogenous promoters; many external chemical agents can act as promoters as well. Eventually, one (or more) of the cells in a premalignant focus sustains a (series of) further genomic event(s) leading to one (or more) malignant cells, and ultimately to a malignant tumor. The process of conversion of an initiated cell into a malignant cell is called malignant conversion.

For the mathematical development, let $X(t)$ represent the number of susceptible cells in the tissue of interest at age t. Then, assume that the number of initiated cells arising from normal susceptible cells is a nonhomogeneous Poisson process with intensity function $f(X(t), v(t))$, where v is a vector of parameters defining the rates of critical genomic events involved in initiation. Suppose that initiated cells divide, die, and become malignant, possibly via stages intermediate between initiation and malignancy, according to a stochastic process. Let $Y(t)$ and $Z(t)$ be random variables that represent, respectively, the number of initiated and malignant cells at time (age) t. Let $\Psi(y, z; t)$ be the probability generating function for the number of initiated and malignant cells at time t. Then the hazard function

$$h(t) = -\Psi'(1, 0; t)/\Psi(1, 0; t)$$

Suppose now that $\Phi(y, z; s, t)$ is the probability generating function for the number of initiated and malignant cells at time t starting with a single initiated cell at time s (i.e., $\Phi(y, z; s, s) = 1$). Then, the process of malignant transformation is a filtered Poisson process and by a generalization of results in Moolgavkar and Venzon (1979) and Moolgavkar and Luebeck (1990)

$$\Psi(y, z; t) = \exp\{\int f(X, v; s)[\Phi(y, z; s, t) - 1]ds\}$$

and thus,

$$h(t) = -\int f(X, v; s)\, \Phi_t(1, 0; s, t)ds$$

where Φ_t is the derivative of Φ with respect to t.

Now, for any t_1 such that $0 < t_1 < t$, we have

$$h(t) = -\int f(X, v; s)\, \Phi_t(1, 0; s, t)ds - \int f(X, v; s)\, \Phi_t(1, 0; s, t)ds$$

It can be easily shown that the first term of this expression $\rightarrow 0$ as $t \rightarrow \infty$ and therefore the asymptotic behavior of the hazard function depends only on the second term. This implies, in particular, that if exposure to an environmental agent modifies some or all of the parameters of the model (thus affecting the hazard function) and if these parameters revert to background levels after exposure to the agent stops, then the hazard function must approach the background hazard (i.e., the hazard in those not exposed to the environmental agent) asymptotically.

The rather general form of the two-mutation clonal expansion model above cannot be used for analyses of data without some further assumptions. It does make it clear, however, that the hazard function of a general multistage model can always be viewed as the hazard function of a two-mutation model with the appropriate waiting time distribution between initiation and malignancy. Thus, without ancillary biological information there is little point to fitting models postulating more than two stages to tumor incidence data. Such information is rarely available, and inferences regarding the number of steps involved in malignant transformation cannot be made from incidence data alone. The simple version of the two-mutation clonal expansion model described below is flexible enough to describe the incidence of tumors in both experimental and epidemiologic studies.

The simple version of the two-mutation model, which has been widely used for analyses of data, makes the following mathematical assumptions. Let $X(s)$ represent the number of normal susceptible cells in the tissue of interest at time (age) s, and suppose that initiated cells arise from normal cells as a nonhomogeneous Poisson process with intensity $v(s)X(s)$, where $v(s)$ is the rate of initiation. Note that v and X are not separately identifiable. However, information on one or the other may be available from independent sources (Moolgavkar and Luebeck 1992). In a small time interval Δs, an initiated cell divides into two initiated cells with probability $\alpha\Delta s + 0(\Delta s)$; it dies or differentiates with probability $\beta\Delta s + 0(\Delta s)$; it divides into one initiated cell and one cell that has sustained the second event (malignant cell) with probability $\mu\Delta s + 0(\Delta s)$. Each of the parameters of the model can be functions of the dose of the environmental agent of interest. The derivation of the hazard, density, probability, and survival functions of the two-mutation model, and their use in the analyses of experimental and epidemiologic data are discussed in various publications (Moolgavkar and Luebeck 1990; Moolgavkar et al. 1993; Luebeck et al. 1996; Kai et al. 1997). It is pointed out here that, although these functions are couched in terms of four parameters (vX, α, β, and μ), there is an identifiability problem in that they depend only on three parameters (Heidenreich 1996). An explicit set of identifiable parameters (Heidenreich 1996; Heidenreich et al., 1997) is given by $M = vX/\alpha$, $P = \alpha(A - 1)$, and $R = \alpha(1 - B)$, where A and B are the roots of the quadratic form associated with the Riccati equation of the model (Moolgavkar and Luebeck 1990). In terms of biological parameters, the latter two parameters (P and R) are approximately $\alpha - \beta$, and $\mu/(1 - \beta/\alpha)$, respectively. In past analyses of data, biological parameters with appropriate constraints have been used. In a recent publication, however, Kai et al. (1997) have used the set of identifiable parameters given above. The advantages of using identifiable parameters are that arbitrary constraints do not have to be imposed on the parameters and convergence of algorithms for likelihood maximization is quick. It is relatively straightforward to construct and maximize the likelihood function in terms of the identifiable parameters.

1. Likelihood Construction and Maximization

Likelihood construction depends upon the study design. Hitherto, biologically based models have not been extensively used for analyses of case-control studies. With cohort studies, when data on individual patterns of exposure are available and each individual in the cohort is followed up for vital status and presence or absence of tumor, the construction of the likelihood depends upon how quickly after its genesis the tumor was diagnosed. Generally, a judgment is made as to whether the tumor was incidental or quickly detected. For an incidental tumor detected at age t, the likelihood contribution is the probability of tumor, $P(t)$, by age t generated by the model under consideration. For a rapidly detected tumor, the contribution is the density function, $P'(t)$. Individuals who are lost to follow-up at age t without developing tumors contribute the survival function, $S(t) = 1 - P(t)$ to the likelihood. The likelihood for the data is the product of contributions made by individual members of the cohort. If information is not available on an individual level, but on the level of fairly homogeneous subgroups, then the likelihood is constructed as in the case of Poisson regression above. In this case the expectation of the Poisson in category i is $N_i h_i(t)$, where N_i represents the person-years at risk in the category and $h_i(t)$ is the hazard function at the midpoint of the age range defining the category. The likelihood of the data is the product of likelihood contributions over all the categories. The likelihood is maximized to estimate the parameters of the model.

2. Examples

Some examples are now given of the use of the two-mutation clonal expansion for analyses of experimental and epidemiologic data. These examples illustrate that specific biological hypotheses can often be addressed with analyses conducted within the framework of this model. The examples given here are of applications of the model to analyses of data on radiation carcinogenesis. In particular, the model is applied to data on radon-induced lung cancer, a topic of considerable importance to the subject matter of this monograph.

a. Radon and Lung Cancer in Rats

The data included in the analysis were from rat experiments conducted under carefully controlled conditions by Dr. Fred Cross at the Battelle Pacific Northwest Laboratories at Richland. The experiments were conducted under radon-daughter exposure conditions that resulted in a dose at the cellular level of approximately 5 mGy per working level month (WLM) of exposure. Data from 1797 animals exposed to radon daughters over the approximate range 320–10,240 WLM (1.1–36 Jhm-3) were included in the analysis. The following information was available on each animal in the data set: the exact age when exposure to radon was begun, the radon-daughter exposure rate in WLM/week (WLM/w), the age at which exposure was stopped, age at death or sacrifice, and presence or absence of malignant lung tumor. All animals were followed until sacrifice or death. The objectives of the analysis were to estimate the mutation rates and intermediate cell proliferation parameters as functions of the exposure rates of radon. This was achieved by maximizing the likelihood of the data. Let $P(t)$ be the probability of tumor by age t for some particular exposure-rate regimen. Then, the survivor function is $S(t) = 1 - P(t)$ and the hazard function $h(t) = -S'(t)/S(t)$. In the opinion of the pathologist, the lung tumors were incidental (i.e., they did not cause death of the animal). Thus, the likelihood of the data was constructed as follows. Because the tumors were incidental, the contribution to the likelihood by an animal that died (or was sacrificed) at age t is $P(t)$ if it had a tumor, or $S(t)$ if it was free of tumor. The full likelihood is the product of these terms over all the animals. The exact expressions and other details can be found in the relevant publication (Moolgavkar et al. 1990). Maximum likelihood estimates of the parameters were obtained. A comparison of observed and expected numbers of tumors in various exposure-rate categories indicated that the model described the data well. Radon was found to increase the first mutation rate and the net proliferation of intermediate cells, but had little effect on the second mutation rate, suggesting that the nature of the two mutational events is different. The analysis also confirmed an inverse exposure-rate effect (i.e., fractionation of a given total exposure to radon increased the risk of lung cancer). Further, the inverse exposure-rate effect could be attributed to the effect of radon on intermediate cell kinetics (i.e., on the promotional effect of radon).

The analysis was extended by Luebeck et al. (1996) to include 3750 rats exposed to varying regimens of radon. New to this analysis was the parameterization of the two-mutation model such that cell killing by α-particles could be explicitly considered. As in the previous analysis, the rate of the first mutation was found to be dependent on radon and consistent with in vitro rates measured experimentally, whereas the rate of the second mutation was not dependent on radon. An initial sharp rise in the net proliferation rate of intermediate cells was found with increasing exposure rate (model I). This model yielded an unrealistically high cell-killing coefficient. A second model (model II) was studied, in which the initial rise was attributed to promotion via a step function, implying that it was not due to radon but to the uranium ore dust that was used as a carrier aerosol. This model resulted in values for the cell-killing coefficient consistent with those found for in vitro cells. An inverse exposure-rate effect was seen attributable, as in the previous analysis, to promotion of intermediate lesions. Since model II is preferable on biological grounds (it yields a plausible cell-killing coefficient), one conclusion of this analysis is that an inverse exposure-rate effect would not be seen in the absence of an irritant such as uranium ore dust.

b. Reanalysis of the Colorado Plateau Uranium Miners' Data

Much of our knowledge regarding the interaction of radon and tobacco smoke in the etiology of human lung cancer derives from studies of uranium miners. A reanalysis of the lung cancer mortality in the Colorado Plateau uranium miners' cohort is briefly described within the framework of the two-mutation model. The analysis takes explicit account of the patterns of exposure to both radon and cigarette smoke experienced by individuals in the cohort. In contrast to the rat lung malignancies, which are incidental, human lung cancers are rapidly fatal. Thus, individuals who develop lung cancer contribute the probability density function for the time-to-tumor to the likelihood function. Individuals who do not develop lung cancer contribute the survivor function, as in the case of the experimental data. The parameters were estimated by maximizing the likelihood. As judged by a comparison of observed and expected number of lung cancers in various categories, the model described the data well. In addition, a comparison of theoretical and empirical Kaplan-Meier plots indicated that the model described the temporal pattern of failures well. A simultaneous reanalysis of the British doctors' cohort indicated that those model parameters relating to the effect of tobacco were similar in the two data sets. No evidence of interaction between radon and cigarette smoke was found with respect to their joint effect on the first or second mutation rates or on the proliferation of intermediate cells. However, the age-specific relative risks associated with joint exposure to radon and cigarette smoke were supraadditive but submultiplicative. The analysis also confirmed an inverse exposure-rate effect (i.e., that fractionation of radon exposure leads to higher lung cancer risks). It is interesting to note that the analyses of experimental and epidemiological data yielded consistent results despite the different likelihoods maximized. Thus, as in the case of the experimental data, analysis of the epidemiological data indicated that radon strongly affected the first mutation rate and the proliferation rate of intermediate cells. It was to the latter effect of radon that the inverse exposure-rate effect could be attributed. If the promotion of intermediate lesions is due to chronic irritation by dust in the mining environment, as the analysis of experimental data suggests (see previous example), then the inverse exposure-rate effect would not be expected with exposure to residential radon. In both epidemiologic and experimental data, the second mutation rate was little affected. For details see Moolgavkar et al. (1993).

c. Cancer Following Exposure to Low LET Radiation

Radon daughters are α-emitters, which is high linear energy transfer (LET) radiation. The model has been applied to γ-radiation (low LET) as well. Little (1995, 1996) has carried out detailed analyses of the A-bomb survivors data using the two-mutation model and extensions of it. He concludes that "without some extra stochastic 'stage' appended (such as might be provided by consideration of the process of development of a malignant clone from a single malignant cell) the two-mutation model is perhaps not well able to describe the pattern of excess risk for solid cancers that is often seen after exposure to radiation." He prefers a three-mutation model for the A-bomb data. Little's analysis is based on consideration of mortality rather than incidence, however. For cancers that are not rapidly fatal, mortality data are a poor surrogate for incidence. The extra stochastic "stage" that Little deems necessary for a satisfactory description of the data could be construed to represent the time between occurrence of the malignant tumor and death. Little (1996) analyzed the incidence of acute lymphocytic leukemia and of chronic lymphocytic leukemia in England and Wales over the period 1971 to 1988, and concluded that the two-mutation model described the incidence of these leukemias well.

Kai et. al (1997) present analyses of the incidence of three solid cancers—lung, stomach, and colon—among the cohort of A-bomb survivors using the two-mutation model. These analyses showed that the temporal evolution of risk following the (essentially) instantaneous exposure to radiation could be explained entirely by the hypothesis that the exposure resulted in the creation of a (dose-dependent) pool of initiated cells that was added to the pool of spontaneously initiated cells. The dose dependence of initiation was consistent with linearity down to the lowest doses in the cohort. There was no evidence of an age dependence of radiation-induced initiation, suggesting that the high excess relative risk seen in those irradiated as children is not due to an inherently higher susceptibility to radiation. Moolgavkar et al. (1997) discuss some implications of these analyses for assessment of radiation risks in other populations and with protraction of exposure. Heidenreich et al. (1997) analyzed the incidence of all solid cancers combined in the cohort of A-bomb survivors using both exact and approximate solutions of the two-mutation model, as well as two empirical models, the "age-at-exposure" model and the "age-attained" model. They concluded that these models, with four parameters estimated for each, described the data well, although the exact two-mutation model described some features of the data better than the other models.

VII. OTHER QUANTITATIVE METHODS

Attention has focused here on the modern statistical methods available for analyses of epidemiologic data, and on biologically based approaches to analyses of experimental and epidemiologic data when the end point of interest is cancer. These methods provide powerful and flexible tools for data and dose–response analyses when information on time-to-disease occurrence is known. More classical methods based on parametric empirical statistical models for time-to-disease occurrence are also sometimes used, particularly in experimental studies. An example of such a model is the Weibull model (Kalbfleisch and Prentice 1980; Cox and Oakes 1984). The hazard function of the Weibull model can be seen to be the approximate form of the hazard function of the Armitage–Doll multistage model (Gart et al. 1986). The empirical statistical models are not well suited to the analyses of time-dependent patterns of exposure or dose. For example, in an inhalation study of fibers and lung cancer in rodents, the time-dependent pattern of lung burden might be directly measured by serial sacrifices during the course of the study. Biologically based carcinogenesis models can be easily used for analyses of such data.

A. Physiologically Based Pharmacokinetic (PBPK) Models

PBPK models provide a powerful tool for the understanding and quantification of the relationship between exposure to an agent and tissue dosimetry. These models attempt to describe the processes that regulate chemical disposition, taking explicit account of the physiological characteristics of the biological system in the species under investigation. In this type of modeling a biological system is envisaged as being comprised of a number of physiologically relevant compartments, and biochemical and physiological parameters are used to describe partitioning of the agent of interest (and its metabolites) among the compartments. The mathematical analysis generally involves the solving of systems of first-order differential equations.

Inhaled toxicants present a special challenge. Not only are PBPK models required to describe the uptake and distribution of inhaled substances, the dynamics of air flow during respiration also play an essential role in determining dose to various parts of the respiratory tree. Recent work on airflow modeling in the respiratory passages holds out the promise of precise estimates of dose of inhaled toxicants to various parts of the respiratory tree (Kimbell et al. 1993). Recent work at the Chemical Industry Institute of Toxicology (CIIT) has exploited airflow, deposition, and uptake models to characterize the quantitative relationship between exposure to formaldehyde and the formation of DNA protein cross-links in the rodent nose. The DNA protein cross-links are used as a biological dosimeter for formaldehyde. Finally, a model of the carcinogenic process, similar to the two-mutation clonal expansion model described above, was used to define the quantitative relationship between DNA protein cross-links and the occurrence of nasal tumors. Using these concatenated series of models, the CIIT investigators were able to explain the pattern of distribution of nasal tumors in rats exposed to formaldehyde and the strong nonlinearity of the tumor response to exposure.

VIII. PROSPECTS FOR THE FUTURE

This chapter has presented a very brief review of the quantitative methods relevant to risk assessment. Improvements in the future will depend on improvements in both the quantitative methodology and in the understanding of underlying biology.

We know that the burden of disease falls unequally on different populations and on different subgroups within the same population, and that genetic, environmental, and life-style factors are all important in determining risk. A central challenge in epidemiology and risk assessment is to quantify the contribution made by each of these to the burden of disease in a population. Until recently, epidemiologic research has not focused specifically on understanding the interaction of genetics and environment in disease. Conventional epidemiology, which has been focused on the study of environmental and life-style factors, often ignored the role of genetic factors or addressed them via a few questions regarding family history. Geneticists, on the other hand, have developed powerful statistical techniques for the identification of major genes involved in disease. Environmental factors have not been incorporated into these analyses and it is clear that epidemiologic studies of the future will need to bridge genetics and environmental epidemiology. Study designs for such investigations that involve combining population-based case-control studies and family studies have been proposed (Whittemore and Gong 1994; Liang and Pulver 1996; Whittemore and Halpern 1997; Zhao et al. 1997).

Exposures are often measured with considerable error in epidemiologic studies. It is well known that exposure measurement errors can lead to bias in estimates of parameters of the statistical model, and distortion of the shape of the exposure–response relationship (Armstrong et al. 1994; Carroll et al. 1995). This issue is of considerable importance in air pollution epidemiology and has received recent attention because the ecologic nature of most epidemiologic studies of air pollution requires that measurements made at central monitoring stations be used as (surrogate and imperfect) measures of personal exposures. The problem is even more complicated when several correlated covariates, some or all of them measured with error, are considered in regression analyses. It is clear that statistical methods to address the effects of measurement error are sorely needed in air pollution epidemiology.

The rapidly developing field of molecular epidemiology offers the hope of assessing dose directly by measurement of biological markers of exposure to specific agents. To date, large intra- and interindividual variations in the quantitative measures of these early biological markers of exposure have limited the usefulness of these methods. There is hope, however, that the discovery of better markers and more precise measurements will lead to improvements. The efforts to develop precise biological markers of exposure complement the development of statistical techniques for addressing problems of exposure measurement error.

Better bioassays need to be developed as well if risks associated with low levels of exposure are to be estimated more precisely. It is not sufficient to record simply the number of animals in each group that develop the response under investigation. The time to response from start of treatment is of equal importance. The quantitative methods briefly described in this chapter can be fully exploited only if such data are available. The design of an "ideal" bioassay is discussed in a recent publication (Moolgavkar et al., in press). A good example of the sort of biological information that can be collected for risk assessment of inhaled toxicants is provided by the recent CIIT risk assessment of formaldehyde, briefly described above. This example illustrates that if a series of models can be developed to describe distinct aspects of the overall disease process, then these models may be concatenated to arrive at an integrated biologically based model of the entire process following exposure to an environmental agent. Connolly and Anderson (1991) give a schema of such a model for carcinogenesis. The construction of such integrated models may be thought of as the Holy Grail of disease process modeling.

BIBLIOGRAPHY

Armstrong, B., White, E., et al. 1994. *Principles of Exposure Measurement in Epidemiology, Monographs on Epidemiology and Biostatistics Vol. 21*, Oxford University Press: New York.

BEIR IV. 1988. *Health Risks of Radon and Other Internally Deposited Alpha-Emitters*, National Academy Press: Washington, DC.

BEIR V. 1990. *Health Effects of Exposure to Low Levels of Ionizing Radiation*, National Academy Press: Washington, D.C.

Benichou J. 1991. Methods of adjustment for estimating the attributable risk in case-control studies: A review, *Statistics in Medicine* 10:1753–1773.

Breslow, N.E., Day, N.E. 1980. *Statistical Methods in Cancer Research, Vol. 1, The Analysis of Case-Control Studies* (IARC Scientific Publications No. 32). IARC: Lyon, France.

Breslow, N.E., Day, N.E. 1987. *Statistical Methods in Cancer Research, Vol. II, The Design and Analysis of Cohort Studies* (IARC Scientific Publications No. 82). IARC: Lyon, France.

Breslow, N.E., Storer, B.E. 1985. General relative risk functions for case-control studies, *American Journal of Epidemiology* 122:149–162.

Bruzzi, P., Green, S.B., et al. 1985. Estimating the population attributable risk for multiple risk factors using case-control data, *American Journal of Epidemiology* 122:904–914.

Burnett, R.T., Dales, R.E., et al. 1994. Effects of low ambient levels of ozone and sulfates on the frequency of respiratory admissions to Ontario hospitals. *Environmental Research* 65:172–194.

Carroll, R.J., Ruppert, D., et al. 1995. Measurement Error in Nonlinear Models. *Monographs on Statistics and Probability*, Chapman and Hall: New York. 63.

Cohen, S.M., Ellwein, L.B. 1990. Cell proliferation in carcinogenesis, *Science* 249:1007–1011.

Connolly, R.B., Andersen, M.E. 1991. Biologically based pharmacodynamic models: Tools for toxicological research and risk assessment, *Annual Review Pharmacology Toxicology* 31:503–523.

Cox, D.R., Oakes, D. 1984. *Analysis of Survival Data, Monographs on Statistics and Applied Probability 21*, Chapman and Hall: New York.

Department of Health and Human Studies. 1994. *Lung Cancer and Radon: A Joint Analysis of 11 Underground Miners Studies* (NIH Publication No 94-3644), U.S. Department of Health and Human Studies, Washington, DC.

Dockery, D., Pope, C., et al. 1993. An association between air pollution and mortality in six U.S. cities, *New England Journal of Medicine*, 329:1753–1759.

English, D.R., Holman, C.D.J., et al. 1995. *The Quantification of Drug-Caused Morbidity and Mortality in Australia, 1995 edition.* Commonwealth Department of Human Services and Health: Canberra, Australia.

Enterline, P.E. 1983. Sorting out multiple causal factors in individual cases, *Methods and Approaches in Occupational and Environmental Epidemiology.* Eds. Chiazze, L., Lundin, F.E., and Watkins, D. Ann Arbor, 177–182.

Gart, J.J., Krewski, D., et al. 1986. *Statistical Methods in Cancer Research, Vol III, The Design and Analysis of Long-Term Animal Experiments* (IARC Scientific Publications No. 79), IARC: Lyon, France.

Greenland, S., Drescher. 1993. Maximum likelihood estimation of the attributable fraction from logistic models, *Biometrics*, 49:865–872.

Greenland, S., Morgenstern, H. 1989. Ecological bias, confounding and effect modification, *International Journal of Epidemiology* 18:269–274.

Greenland, S., Robins, J. 1986. Identifiability, exchangeability, and epidemiological confounding, *International Journal of Epidemiology* 15:413–419.

Greenland, S., Robins, J. 1994. Invited commentary: Ecologic studies — biases, misconceptions and counterexamples, *American Journal of Epidemiology* 139:747–759.

Guerrero, V.M., Johnson, R.A. 1982. Use of the Box-Cox transformation with binary response models, *Biometrika* 69:309–314.

Health Effects Institute (HEI). 1995. *Particulate Air Pollution and Daily Mortality: Replication and Validation of Selected Studies, The Phase I Report of the Particle Epidemiology Evaluation Project.* Health Effects Institute: Cambridge, MA.

Heidenreich, W.F. 1996. On the parameters of the clonal exapnsion model, *Radiation and Environmental Biophysics* 35:127–129.

Heidenreich, W., Luebeck, E.G., et al. 1997. Some properties of the hazard function of the two-mutation clonal expansion model, *Risk Analysis* 17:391–399.

Kai, M., Luebeck, E.G., et al. 1997. Analysis of solid cancer incidence among atomic bomb survivors using a two-stage model of carcinogenesis, *Radiation Research* 148:348–358.

Kalbfleisch, J.D., Prentice, R.L. 1980. *The Statistical Analysis of Failure Time Data*, Wiley Series in Probability and Mathematical Statistics, John Wiley & Sons: New York.

Kimbell, J.S., Gross, E.A., et al. 1993. Application of computation fluid dynamics to regional dosimetry of inhaled chemicals in the upper respiratory tract of the rat, *Toxicology and Applied Pharmacology* 121:253–263.

Liang, K.Y., Pulver, A.E. 1996. Analysis of case-control/family sampling design, *Genet. Epidemiol.* 13:253–270.

Little, M.P. 1995. Are two mutations sufficient to cause cancer? Some generalization to the two-mutation model of carcinogenesis of Moolgavkar, Venzon, and Knudson, and of the multistage model of Armitage and Doll, *Biometrics* 51:1278–1291.

Little, M.P. 1996. Generalization of the two-mutation and classical multistage models of carcinogenesis fitted to the Japanese atomic bomb survivor data,*J. Radiol. Prot.*, 16:7–24.

Luebeck, E.G., Curtis, S.B., et al. 1996. Two-stage model of radon-induced malignant lung tumors in rats: Effects of cell killing, *Radiation Research* 145:163–173.

Luebeck, E.G., Moolgavkar, S.H. 1994. Simulating the process of carcinogenesis. *Mathematical Biosciences* 123:127–146.

McCullagh, P., Nelder, J.A. 1989. *Generalized Linear Models, 2nd edition*, Chapman and Hall: New York.

Moolgavkar, S.H. 1978. The multistage theory of carcinogenesis and the age distribution of cancer in man, *JNCI* 61:49–52.

Moolgavkar, S.H. 1991. Stochastic models of carcinogenesis, *Handbook of Statistics*, Vol. 8. In: Rao, C.R., Chakraborty R, Eds., Elsevier: New York. 373.

Moolgavkar, S.H., Cross, F.T., et al. 1990. A two-mutation model for radon-induced lung tumors in rats, *Radiation Research* 121:28–37.

Moolgavkar, S.H., Dewanji, A., et al. 1989. Cigarette smoking and lung cancer: Reanalysis of the British doctors' data, *Journal of the National Cancer Institute* 81:415–420.

Moolgavkar, S.H., Krewski, D., et al. (In press). Mechanisms of carcinogenesis and biologically-based models for quantitative estimation and prediction of cancer risk. In: *Quantitative Estimation and Prediction of Cancer Risk*, Moolgavkar, S.H., Krewski, D., Zeise, L., Cardis, E., Moller, H., Eds., IARC Scientific Publications, 131.

Moolgavkar, S.H., Luebeck, E.G. 1990. Two-event model for carcinogenesis: Biological, mathematical and statistical considerations, *Risk Analysis* 10:323–341.

Moolgavkar, S.H., Luebeck, E.G. 1992. Multistage carcinogenesis: A population-based model for colon cancer, *JNCI* 84:610–618.

Moolgavkar, S.H., Luebeck, E.G. 1996. Particulate air pollution and mortality: A critical review of evidence, *Epidemiology* 7:420–428.

Moolgavkar, S.H., Luebeck, E.G., et al. 1993. Radon, cigarette smoke, and lung cancer: A reanalysis of the Colorado Plateau miners' data, *Epidemiology* 4:204–217.

Moolgavkar, S.H., Luebeck, E.G., et al. 1997. Air pollution and hospital admissions for respiratory causes in Minneapolis-St. Paul and Birmingham, *Epidemiology* 8(4):364–370.

Moolgavkar, S.H., Venzon, D.J. 1979. Two-event model for carcinogenesis: Incidence curves for childhood and adult tumors, *Mathematical Biosciences* 47:55–77.

Moolgavkar, S.H., Venzon, D.J. 1987. General relative risk models for epidemiologic studies, *American Journal of Epidemiology* 126:949–961.

Pope, C.A. III, Thun, M.J., et al. 1995. Particulate air pollution as a predictor of mortality in a prospective study of U.S. Adults. *American Journal of Respir. Crit. Care Med.* 151:669–674.

Prentice, R.L., Moolgavkar, S.H., et al. 1986. Biostatistical issues and concepts in epidemiologic research, *Journal of Chronic Diseases* 38:1169–1183.

Rothman, K.J. 1986. *Modern Epidemiology.* Little, Brown & Co.: Boston, MA.

Rothman, K.J., Greenland, S. 1998. *Modern Epidemiology, 2nd Edition,* Lippincott-Raven: New York.

Schwartz, J. 1993. Air pollution and daily mortality in Birmingham, Alabama, *American Journal of Epidemiology* 137:1136–1147.

Schwartz, J., Dockery, D.W. 1992. Increased mortality in Philadelphia associated with daily air pollution concentrations, *American Review of Respiratory Disease* 145:600–604.

Tan, W.Y. 1991. *Stochastic Models of Carcinogenesis,* Marcel Dekker: New York.

Thomas, D.C. 1981. General relative risk models for survival time and matched case-control analysis, *Biometrics* 37:673–686.

Venzon, D.J., Moolgavkar, S.H. 1988. Origin invariant relative risk functions for case-control and survival studies, *Biometrika* 75:325–333.

Wahrendorf, J. 1987. An estimate of the proportion of colo-rectal and stomach cancers which might be prevented by certain changes in dietary habits, *International Journal of Cancer* 40:625–628.

Whittemore, A.S., Gong, G. 1994. Segregation analysis of case-control data using generalized estimating equations, *Biometrics* 50:1073–1087.

Whittemore, A.S., Halpern, J. 1997. Multiphase sampling designs in genetic epidemiology, *Stat. Med.* 16:153–167.

Yang, G.C., Chen, C.W. 1991. A stochastic two-stage carcinogenesis model: A new approach to computing the probability of observing tumor in animal bioassays, *Mathematical Biosciences* 104:247–258.

Zhao, L.P., Hsu, L., et al. 1997. Population-based family study designs: An interdisciplinary research framework for genetic epidemiology, *Genet. Epidemiol.* 14:365–388.

CHAPTER 5

Exposure Characterization

David R. Patrick

CONTENTS

I. INTRODUCTION

The National Research Council (NRC 1991) described human exposure to a contaminant as an event consisting of contact with a specific contaminant concentration

1-56670-323-9/99/$0.00+$.50

at a boundary between the human and the environment (e.g., lung or skin) for a specific interval. Total exposure is determined by multiplying the concentration by the exposure time. Exposure is translated into a biologically effective dose as some or all of the contaminant is absorbed or deposited in the body, a process that can depend upon numerous factors including chemical and physical properties of the contaminant, mode of entry into the body, breathing rate, and metabolic factors. As such, an exposure assessment can require evaluation of some or all of the following: sources; environmental media through which exposure occurs; transport from the source to the receptor; chemical and physical transformations; routes of entry to the body; intensity and frequency of contact; and spatial and temporal concentration patterns.

There are three basic methods for estimating human exposure to an environmental contaminant. The first two are *direct* measures of exposure while the third is an *indirect* measure of exposure.

1. A person can wear a device that periodically or continuously measures, at or near a likely site of entry, the concentration of the contaminant(s) of concern. This is usually the most accurate method but is also expensive and time consuming.
2. Exposure can be estimated from the contaminant's actual dose in the body if it can be measured or if it manifests itself in a measurable way (e.g., in the urine or as a metabolite in the bloodstream). These biomarkers are less widely used because they generally require medical evaluation and, in some cases, invasive testing, and they require considerable knowledge of the physical or biological processes of the body.
3. Exposure can be inferred by measuring contaminant concentrations in the environment (indoors or outdoors) to which the person can be exposed and then estimating the internal dose by using scientifically accepted exposure factors or calculation methods. This method is used most widely because it can be applied relatively easily to large populations and large geographic areas.

Public health officials typically characterize environmental exposure and subsequent risks by investigating several populations, including:

1. all individuals potentially exposed to a contaminant (i.e., the exposed population);
2. the one or more individuals who are exposed to the contaminant to the greatest extent (i.e., the most exposed, or maximally exposed, individual); and
3. persons who may be particularly sensitive to one or more contaminants (e.g., children, the elderly, the ill, or the infirm).

While estimation of the total population exposed to the contaminant is relatively straightforward, the determination of the most exposed individual has been a source of controversy. A primary reason for the controversy is that a number of assumptions generally are required to define the most exposed individual and there is often disagreement on these assumptions. For example, some assessments consider the most exposed individual to be the person exposed to the maximum ambient concentration of a contaminant, calculated using worst-case emission and dispersion assumptions and assuming continuous exposure for a lifetime (usually 70 years). While this maximum exposure is theoretically possible, it is almost always unrealistic. Improved exposure assessments use more advanced mathematical techniques and data, such as statistical distributions, for describing realistic maximum as well as actual exposures. Regulators are also moving away from use of ambiguous terms like "maximally exposed individual," in recognition of the difficulty in agreeing upon their meaning. Identification of "sensitive individuals" can also be controversial for many reasons including the difficulty in assigning specific exposures to specific adverse effects and because sensitivity can be associated with a wide variety of physical and genetic factors as well as psychological reactions. Children, the elderly, and the infirm are clearly groups of special concern. Moreover, some investigators currently hypothesize that exposures to low levels of chemical mixtures indoors, or to some common indoor air pollutants, may be associated with identifiable adverse effects in some otherwise healthy individuals.

In the past, legislators and regulators separately treated human exposures resulting from contact with different environmental media (i.e., air, water, and waste materials). As such, potential exposures through inhalation, ingestion, and skin contact were usually evaluated independently. This occurred largely because the different media were separate and most research focused on one media. Today, we know that these media are often interconnected and that some air pollutants, for example, can deposit onto and contaminate water bodies, the earth's surface, and plant and animal life. These different media exposures are being combined more frequently in multipathway (meaning all likely routes) exposure and risk assessments to approximate more closely actual exposures and risks.

The purpose of this chapter is to describe the process of exposure assessment, the methods used to conduct such assessments, and the application of these methods to indoor air analyses. Exposure assessment can involve a variety of physical calculations and computerized methods and techniques; this chapter identifies the most widely used methods and techniques and describes the more important advantages and disadvantages.

II. IMPORTANT EXPOSURE ASSESSMENT CONCEPTS

Several concepts are important to properly conduct and understand exposure assessments. As described in EPA (1992), the process of a chemical entering the body occurs in three basic steps:

1. the human comes into contact with, or is exposed to, a chemical in the air, water, food, and soil;
2. an amount of the chemical crosses a boundary from outside to inside the body, through intake (e.g., inhalation or ingestion) or uptake (e.g., absorption through the skin), and subsequently is absorbed and becomes available at biologically significant sites; and
3. an amount of the chemical reaches a target site and results in an adverse effect.

This process gives rise to several concepts of dose. The *applied dose* is the amount of the chemical in contact with the barrier (i.e., lung, gastrointestinal tract, or skin) that is available for absorption. The *potential dose* is the amount of the chemical that is inhaled, ingested, or applied to the skin. The *internal dose*, also called the *absorbed dose*, is the amount of the chemical or its product that is absorbed and is available for interaction with biologically significant receptors. Once absorbed, the chemical can undergo metabolism, storage, excretion, or transport within the body. The amount transported to the organ, tissue, or fluid of interest is called the *delivered dose*. Finally, the *biologically effective dose* is the amount that actually reaches cells, sites, and membranes where it gives rise to adverse effects. In most instances, the indoor exposure and risk assessment will focus on the applied or potential dose because consideration of the internal, delivered, and biologically effective doses requires an understanding of human biological and chemical processes. These latter dose concepts are important to scientists attempting to develop acceptable health criteria for the range of possible chemical exposures.

Exposure and dose can be estimated in various ways. Exposure concentrations are useful when comparing peak exposures to health criteria such as the OSHA short-term exposure limits (STEL). Time-weighted averages are widely used by the OSHA for work-day occupational exposures and by the EPA in conducting carcinogen risk assessments. Exposure or dose profiles describe concentration or dose as a function of time and can be important where both concentration and time are important. Finally, integrated exposures can be useful where the total exposure rather than the exposure profile is important.

As indicated earlier, exposure can be estimated in three different ways.

1. Exposure can be estimated at the point of contact by measuring both exposure concentration and time of contact.
2. Exposure can be estimated by separately evaluating the exposure concentration and the time of contact and then combining the information.
3. Exposure can be estimated from dose, which is determined through biomarkers, excretion levels, or other means after the exposure has taken place.

Exposure and dose information that appropriately estimates the important risks must also be gathered. *Individual risk* is frequently estimated and is the risk borne by a person or group of persons in the population. In the past, regulators often focused on the maximum exposed individual in calculating the individual risk, although the definition of maximum varied. For example, the concept of *maximum* changes significantly depending upon the use of modeled or measured data and actual or theoretical exposure, and consideration of exposure location, special sensitivities (e.g., children, gender, the elderly, or the infirm), and whether point estimates or probability distributions are used.

The EPA is generally moving away from the use of the term "maximally exposed individual (MEI)" because of the difficulties in agreeing on the above factors. In the exposure assessment guidelines (EPA 1992), the EPA described two other terms for consideration in place of MEI:

High-end exposure estimate (HEEE) — A plausible estimate of the individual exposure of those persons at the upper end of the exposure distribution. High-end is stated conceptually as above the 90th percentile of the population distribution, but not higher than the individual in the population who has the highest exposure.

Theoretical upper-bounding estimate (TUBE) — A bounding value that is easily calculated and is designed to estimate exposure, dose, and risk levels that are expected to exceed the levels experienced by all individuals in the actual distribution. The TUBE is calculated by assuming limits for all variables used to calculate exposure and dose that, when combined, will result in mathematically highest exposure or dose.

Population risk is also important. Population risk is the estimate of the extent of harm to the total exposed population. Population exposure and risk can include: the portion of the population that exceeds an accepted health criteria or is within a specified risk range; the exposure or risk to a particular population subgroup; probabilistic estimates; and exposures or risks averaged over specified times (e.g., a year). In carcinogen risk assessments, the EPA often considers the following two population risks:

1. *Risk distribution* — The distribution of individual risk across the exposed population (i.e., the number of individuals in various risk intervals, such as between 10^{-4} and 10^{-5} or 10^{-5} and 10^{-6}). This is calculated by combining the population distribution with the concentration distribution within a specified distance of the source of emissions.

2. *Average annual incidence* — A point estimate of the total population risk. This is estimated by multiplying the number of people at each risk interval by that risk and totaling the estimated number of lifetime cancer deaths. For example, if ten people are exposed to a carcinogen at a risk level of one in ten, one cancer death would be estimated. Since cancer risk estimates are for a 70-year lifetime, the average annual incidence is determined by dividing the total by 70.

The exposure assessment is intended primarily to estimate a dose which is combined with dose–response data to estimate risk. However, exposure assessments can support an array of decisions ranging from priority setting to regulatory control. The end use of the exposure assessment dictates the quality and quantity of information used. Regulatory control decisions typically require higher quality and more detailed information than priority setting decisions because greater societal cost is potentially involved. Regulatory control decisions also require that the link between the source and the exposed or potentially exposed population be established more accurately. Exposure assessment for screening purposes and priority setting can often focus on comparative exposures and risks, with estimates often presented in broad categories (e.g., high, medium, and low). The important rule to remember is that the scope, depth, and cost of the investigation should be determined by the ultimate purpose for the exposure assessment.

III. THE COMPONENTS OF THE INDOOR AIR
EXPOSURE ASSESSMENT

The EPA's Guidelines for Exposure Assessment (EPA 1992) identify five principal components of a typical exposure assessment:

Sources and pollutants — The pollutants and their relevant sources in the environment must be identified, including production, use, disposal, and environmental pathways.

Exposure pathways and environmental fate — The ways in which the pollutant reaches the exposed individual or population (i.e., the receptor), including the movement through and any changes in the environment, must be determined and analyzed.

Measured or estimated concentrations — The environmental concentrations of the substance that are available for exposure must be determined based on measured data, use of mathematical models, or both.

Exposed populations — Populations, particularly sensitive populations, that are potentially exposed by various routes of interest must be identified.

Integrated exposure analysis — The integrated exposure analysis generally combines the estimation of environmental concentrations with the description of the exposed population to yield exposure profiles. For many analyses, the results should be considered in conjunction with the geographical distribution of the human or environmental populations.

Exposures can occur in several different indoor environments, called microenvironments, including at home and work, in transit, and in other indoor locations. These exposures should be estimated in ways that facilitate ready integration with the dose–response assessment data to allow estimation of risk. In addition, information for each of the five principal areas listed above may be limited for scientific, resource, or other means. The exposure assessor must evaluate the information and its limitations and, as noted by NRC (1991), determine how accurately the exposure or exposure potential estimate must be in order to facilitate appropriate risk assessment and risk management decisions.

The following sections provide a more thorough description of the above five components as applied to indoor air exposure assessments.

A. Identifying Pollutants and Sources of Indoor Air Contaminants

In the Report to Congress on Indoor Air Quality (EPA 1989), the EPA grouped indoor air pollutants of concern into the following broad categories, although some of these categories overlap:

Environmental tobacco smoke (ETS) — Includes smoke from the end of the cigarette, cigar, or pipe and smoke exhaled by the smokers. The primary sources are smokers in the indoor area of concern and nearby outdoor sources. ETS includes volatile organic compounds, formaldehyde, polycyclic organic matter, and particulate matter.

Radon and radon daughters — Colorless, odorless, radioactive gases that are decay products from some widely occurring rock formations. The primary sources are underlying soil, well water, and some building materials.

Biological contaminants — Includes molds, pollen, bacteria, viruses, insect and arachnid excreta, and animal and human dander. There are numerous indoor and outdoor sources of biological contaminants.

Volatile organic compounds (VOCs) — This class of pollutants can be large depending upon the definition.[1] An organic compound is any compound of carbon, excluding carbon monoxide, carbon dioxide, carbonic acid, metallic carbides or carbonates, and ammonium carbonate. Organic compounds that are realistically volatile enough to be emitted into the air are usually those with a limited number of carbon atoms (some restrict VOCs to organic compounds with 12 or less carbon atoms). Organic compounds that are emitted as particles or are adsorbed onto particles generally are not available for photochemical reactions or for gas-phase reactions but can be taken into the body by inhalation, ingestion, or skin contact. Important sources of VOCs include paints, stains, adhesives, dyes, solvents, caulks, cleaners, pesticides, building materials, office equipment, and petroleum products.

Formaldehyde — Although technically a VOC, formaldehyde frequently is considered separately. Important sources of formaldehyde include ETS, some foam insulations, particle board, plywood, furnishings, and upholstery.

Polycyclic organic compounds — This class of compounds also can be large. Polycyclic organic matter (POM) is defined in the 1990 Clean Air Act Amendments as organic compounds with more than one benzene ring and that have a boiling point equal to or greater than 100°C. POM can include substances with atoms other than carbon and hydrogen (e.g., oxygen, chlorine, or nitrogen); polycyclic aromatic hydrocarbons (PAH) are a subset of POM containing only carbon and hydrogen atoms. Important sources of POM include combustion processes, particularly incomplete combustion sources, and pesticides application.

Pesticides — There are a large number of chemicals used worldwide as pesticides. Pesticides can be applied professionally or individually both inside and outside buildings and structures. Exposure can occur during application through inhalation of aerosols and gases, and by subsequently breathing emissions, contacting surfaces upon which the pesticide is applied, or by consuming solids and liquids contaminated with pesticides.

Asbestos — Once widely used as a fire retardant and insulation, new asbestos use has all but disappeared in the U.S. since the early 1970s. However, many buildings and homes built before that time still contain asbestos. Asbestos is usually of little concern if intact, but of greater concern if friable (i.e., breaking down and releasing its small fibers) and during demolition.

Combustion products — The combustion of fuels used in human endeavors gives rise to several waste products. The combustion products of primary concern are carbon monoxide, carbon dioxide, nitrogen oxides, sulfur oxides, and particulate matter. Moreover, VOCs, formaldehyde, POM, trace metals, and residues of chemicals in the fuels can also be released during combustion.

---

[1] U.S. regulators often define VOCs differently for different programs. For example, urban smog (i.e., tropospheric ozone) is formed in a photochemical reaction between VOCs and nitrogen oxides. For this program, the EPA defines VOCs as any organic compound except for a relatively small number that are specifically excluded because they are considered not to be photochemically reactive.

Particulate matter — Particulate matter is defined by the EPA as "any finely divided solid or liquid material, other than uncombined water . . ." (40 CFR 60.2). Particulate matter arises from natural and manmade sources and can exist in a wide range of particle sizes and compositions. The size of the particle determines in large part how it affects humans. Larger particles cause soiling and humans can come into contact with them through skin contact and inhalation, and through ingestion of contaminated foods and liquids. However, larger particles are generally captured in the mouth, nose, and upper respiratory tract when inhaled. Human exposure to smaller particles is similar to large particles except that small particles are taken more deeply into the lungs; inhalable particles are defined by the EPA as equal to or less than 10 microns[2] in diameter (40 CFR 50.6). Particles generally less than about 2 microns in diameter also principally affect optical visibility. Particles less than 1 micron in diameter become increasingly difficult to distinguish from vapors.

An indoor air assessment is typically instigated by an adverse health effect that may or may not be attributable to a single or multiple contaminants. If the contaminant is known, the source may already be known and the unwanted release can be contained or mitigated. However, when the adverse effect cannot be attributed to a particular contaminant, sources and pollutants must be postulated and evaluated. Studies show that common indoor sources of contaminants are combustion appliances and equipment, consumer and commercial products, building materials and furnishings, pesticides, heating-ventilation-air conditioning (HVAC) systems, water-damaged materials, humans and pets, and personal activities such as cooking and smoking. Typical indoor pollutants also enter with the ambient outside air, water supplies, and nearby soil. These can be increased near industrial, commercial, or public activities. The relationships between the contaminants and their sources can be complex and a single contaminant can result from several sources both indoors and outdoors (EPA 1989).

The determination of the pollutants and the sources in the indoor environment requires a knowledge of the building, the building occupants, and the surrounding sources of pollutants. Building location, design, operation, maintenance, age, and other factors can substantially affect the concentrations of pollutants; occupants (both permanent and transient) can give rise to pollutants and bring pollutants into a building; and pollutants emitted into the surrounding ambient air and those in the water (groundwater and drinking water) and soil can enter the building and even be concentrated in some situations. While some indoor pollutants and sources can be inferred from surveys or by using models, for maximum utility in the exposure assessment this information should be obtained by observation and analysis whenever possible.

Pollutants and sources in a building often are determined through measurement. The emissions of pollutants from materials used indoors can be measured through laboratory tests, often in enclosed chambers under carefully controlled conditions. A number of laboratories in the U.S. have test chambers in which materials are tested. Four of the more prominent are the following:

---

[2] One micron is one millionth of one meter.

1. EPA's Air and Energy Engineering Research Laboratory, Research Triangle Park, North Carolina
2. Oak Ridge National Laboratory, Oak Ridge, Tennessee
3. Lawrence Berkeley Laboratory, Berkeley, California
4. Georgia Tech Research Institute, Atlanta, Georgia

Models are also used to identify pollutants and sources when direct measurement is technically or economically infeasible. In addition, models are particularly useful in assessing differences in emissions resulting from changes in material, design, or operation; this function could be much more costly if undertaken through measurement. Models can also address past or future situations that cannot be measured directly.

B. Determining Exposure Pathways and Environmental Fate

The magnitude of human exposure to a contaminant depends on the concentration of the contaminant, how the person comes into contact with the contaminant, and the time of exposure. The next step in estimating indoor exposure to a pollutant or pollutants is to identify the exposure pathways (i.e., the routes a chemical takes from the source to the person) and then to determine whether the pollutants change between release and intake (i.e., the environmental fate). Importantly, for exposure to occur all of the following must be present:

- contaminant source,
- transport medium (e.g., air, water, or soil),
- point of contact with the contaminated medium,
- exposure route (e.g., inhalation, ingestion, or skin contact),
- receptor.

Outdoor exposure pathways can be complex and involve varieties of contaminants, contaminant sources, transport media, points of contact, and exposure routes. Indoor exposure pathways are generally less complicated, often involving fewer transport media, points of contact, and exposure routes. Still, a full exposure pathway analysis of indoor air can be complex.

Typical pathways of the substances to which indoor populations are exposed are infiltration of contaminated outdoor environment (in air, water, or soil), contaminants brought indoors from the outside by the occupants, volatilization and evaporation of chemicals from indoor surfaces, emissions from indoor equipment (e.g., furnaces), indoor spaces with a potential for emissions (e.g., garages), personal activities such as cooking and smoking, and many others. Proper delineation of the sources and substances can require an understanding of the chemical and physical properties, use of mass transfer information to estimate movement, and possibly investigation using hydrogeology, soil characterization, topography, and meteorology. Once released, many pollutants change as a result of chemical, physical, or biological processes in the atmosphere, water, or soil. While this occurs less frequently in a stable indoor environment, which is often characterized by relatively limited changes in physical and chemical conditions, pollutants that infiltrate from outside may have changed in form or nature after release and before infiltration. In addition, particulate matter can settle gravimetrically indoors and many gases and vapors can be adsorbed or absorbed onto indoor surfaces. Contaminant concentrations can increase, decrease, or remain constant in an indoor environment in a dynamic process that depends upon such factors as the sources, rates of release, ventilation, and air exchanges.

The assessment of exposure pathways and environmental fate typically requires a combination of theoretical analysis and measurement. The theoretical analysis is often necessary to postulate the sources and the pollutants; measurement can then confirm the presence or absence of the postulated pollutant or source. Exposure pathways are somewhat more limited indoors than outdoors. The air present for inhalation is usually more consistent in composition and character than outdoor air but still is influenced by many factors. Exposure from water sources occurs predominantly from ingestion of drinking and cooking water, and inhalation of volatilized organics and radon from water being used (e.g., in showers). The soil is predominantly a pathway for indoor exposures to radon and chemicals such as pesticides. In the outdoor environment, environmental fate can be an important factor in determining the precise pollutants and concentrations to which people are exposed. For example, many substances are chemically altered in the air, water, and soil; many volatile organic chemicals react in the presence of nitrogen oxides and sunlight to form ozone and other photochemical oxidants; and some otherwise innocuous chemicals can react with other chemicals or degrade to form toxic chemicals. In the indoor environment, these processes are lessened although not eliminated. For example, particulate matter indoors can settle gravimetrically and change from an inhalation concern to a skin contact and ingestion concern (e.g., children crawling on floors get dust on their hands, and then put their hands in their mouths).

C. Measuring or Estimating Indoor Air Concentrations

Indoor concentrations are measured or estimated. The choice of method is dictated by such factors as the pollutant, the sources, the breadth of the area or population under consideration, the use of the information, and the cost. Direct measurements can be taken by using personal monitors and by determining the presence of biological markers in the exposed population. Personal monitoring involves direct measurement of concentrations of air contaminants, generally in the breathing zone of an individual. Indoor concentrations can also be measured indirectly using fixed or portable monitors and by testing the equipment (e.g., HVAC ducts) or materials (e.g., water in chillers) suspected of contributing to the indoor air pollutant concentrations of concern. Monitors are usually classified as active (i.e., relying on a pump or blower to collect samples) or passive (i.e., relying on diffusion to collect samples). Chemical analysis in the laboratory predominates because real-time instrumental analyzers are often large, complex, and expensive, particularly when more than one pollutant is being measured.

Measurement protocols are developed by regulatory agencies, independent organizations, industrial firms, and others. Generally, however, regulatory agency procedures must be complied with in order to ensure regulatory acceptance of the test results. Biological contaminants are among the most difficult to measure and quantify. Current techniques typically involve collection of air and/or surface samples, the culturing of the biological particles, and microscopic counting and identification of the biological entities. In some instances, further biochemical or immunological analysis may be required.

The use of direct measurement methods is limited in several ways.

1. Concentrations of indoor air contaminants are often too low for current methods to measure accurately and reproducibly unless sampling is carried out over a long period of time.
2. Normal background levels, particularly of biological contaminants, are not well understood, making interpretation of the results difficult.
3. Concentrations of many indoor contaminants can vary significantly across microenvironments (e.g., from room to room in a residence).
4. Some chemical compounds interfere with the measurement of other chemical compounds, particularly at low concentrations.
5. The cost of direct measurement can be high; this often limits scope and applicability.

As described by the National Research Council (NRC 1991), the use of personal monitors in or near the breathing zone is the direct method most often used to measure a specific individual's exposure to a contaminant or group of contaminants. Typically, these monitors are carried for a few days to ensure that enough sample is obtained for analysis. These methods have been used extensively for many years by industrial hygienists and others to study occupational exposures. This type of monitoring can also provide an integrated sample across the microenvironments in which the subject moves during the course of the sample period, including the home, in transit, and at work. If information on the person's physical movement is needed, it is often obtained through diaries maintained by the subject. The physical environment, including ventilation, temperature, and humidity, is typically monitored separately. While broadly useful, personal sampling is limited in several ways.

1. People often do not want to be bothered with the inconvenience of carrying a personal sampler and maintaining a diary of activities.
2. Personal samplers necessarily are small and lightweight. This limits their complexity and sensitivity, although recent advances have led to improved equipment.
3. Adequate test design requires a relatively large number of subjects. Maintaining and monitoring a large test population can be resource intensive and time-consuming.

Biological markers are chemical or physical changes in exposed persons that are the direct result of exposure to one or more specific air contaminants. When there is scientific evidence that exposure to a specific substance can be measured biologically or that it gives rise to other substances that can be measured biologically, biological markers are a particularly valuable means of confirming previous exposures to specific substances. Some common biological markers are: cotinine in the urine that results from nicotine exposure; carboxyhemoglobin in the blood that can result from exposure to carbon monoxide and at excessive levels can fatally reduce the ability of the blood to process oxygen; and lead in blood, teeth, and hair that results from exposure to lead through inhalation and ingestion. The use of biological markers, however, is also limited in several ways.

1. Some information must be gathered by invasive means (e.g., blood tests) which can be difficult to obtain.
2. The scientific understanding of the relationship between biological marker concentrations and exposure is frequently not well understood.
3. Biological test results can vary considerably from person to person.
4. Biological testing can be expensive and generally requires highly trained test and analytical personnel.

Indirect methods of measurement often are preferred for estimating exposure because they can be used to reasonably describe an individual's exposure and they are generally less costly. The personal monitoring methods are often combined with indirect measurements in the microenvironment or modeling with information gathered from or about the exposed population. Indirect methods offer advantages that include being able to sample and analyze pollutants in real time. This is in contrast to many direct measurements which usually take samples for subsequent analysis. However, indirect measurement has limitations.

1. Most indirect measurement methods are stationary so that they measure the pollutant concentrations in a specific microenvironment rather than the concentrations a human encounters as he or she moves through microenvironments.
2. Outdoor air pollutant concentrations are monitored continuously by a nationwide system of federal, state, and local monitors. For reasons including convenience and cost reduction, outdoor monitoring results often are used as surrogates for indoor air concentrations. Comparative studies show that indoor exposure estimates based on outdoor monitoring typically differ significantly from estimates based on indoor monitoring, and the degree of difference varies between pollutants.
3. Indirect measurement methods that sample and analyze specific pollutants are often bulky and expensive.
4. In comparisons, indirect monitoring usually predicts higher exposures than personal monitoring.

In order to maximize the accuracy and reproducibility of the measurement results, a monitoring protocol should be developed and followed. A number of studies in the U.S. in recent years were designed to measure concentrations of air contaminants indoors. Various collection and analytical methods were used and some of these studies combined both measurement and modeling. Several of the studies that are most useful in understanding the indoor environment are summarized in Chapter 8. Importantly, the EPA prepared a compendium of indoor air test methods (EPA 1987). This compendium provides available and accepted protocols for measuring selected indoor air pollutants. In some cases, however, there are no generally accepted protocols.

Models are also used to estimate microenvironment concentrations and exposures. Numerous mathematical models have been developed and find particular utility when a larger population or number of indoor microenvironments must be investigated. In these instances, direct and indirect measurements are often prohibitively expensive. Moreover, measurement is not possible for a prospective source or to estimate exposures retrospectively for an epidemiology study. In these and many other cases, models must be used.

Mathematical models are usually classified into two main categories: (1) models that predict concentration, and (2) models that predict exposure. Models are typically derived from fundamental physical and chemical relationships and can focus on individuals or populations. In an individual exposure model, microenvironment contaminant concentrations are measured or modeled and time-activity patterns are used to estimate the time spent in each. Total exposure then is determined by summing the products of concentration and time for all of the microenvironments in which the individual spent time. In a population exposure model, the microenvironment concentrations are combined with individual activity patterns and the results extrapolated to a larger population. Many models assess relatively small populations by taking into account activity patterns and different types of exposure; however, it becomes increasingly costly and complex for a model to attempt to deal with a large number of specific individuals who may be uniquely exposed. While models are widely used, there are often difficulties in validating the results and uncertainties in the mathematical expression of some activities and events. In addition, many models use data from source emission testing and field monitoring to calibrate and verify the components of the model.

The EPA's Report to Congress on Indoor Air describes four general categories of indoor air models.

Source emission models are used to predict emissions from indoor sources of pollutants. For example, models are available to: (1) estimate emission factors from combustion sources that attempt to account for variability in age, condition, and use patterns; (2) estimate emissions of formaldehyde from particle board; and (3) estimate the effects of population activities and activity patterns on indoor pollutant concentrations.

Indoor air quality models, also called transport models, attempt to characterize the movement of air pollutants through defined indoor spaces and estimate the pollutant concentrations under specified conditions. These models investigate the physical pathways through which air is moved within the indoor environment and can include consideration of infiltration through building openings, movement through internal passageways, movement through ventilation and heating ductwork, and the effects of wind and thermal buoyancy. The National Bureau of Standards (NBS) (now the National Institute for Science and Technology, or NIST) developed such a model that accounted for pollutant generation, dilution, reaction, and removal as well as infiltration and exfiltration (EPA 1989). The EPA's Air and Energy Engineering Research Laboratory also developed a preliminary version of an indoor air quality model which used a basic mass balance equation and could include sources such as cigarette smoking, kerosene stoves, and unvented stoves.

Statistical models allow researchers to expand the results of field studies to larger populations. These models use empirical data gathered on variables such as pollutant concentrations, building volumes, and air flow patterns. Increasingly, Monte Carlo computer simulations are being used to describe the statistical distributions.

Population exposure models are available that estimate indoor and outdoor air exposures and allow investigation of a wide range of conditions. These models typically incorporate for the subject population input data on the pollutant concentrations and route of exposure, time-activity patterns, and often health or demographic characteristics that affect exposure.

Three of the better known models used for indoor air studies are the following:

SHAPE (Simulation of Human Air Pollution Exposure) — This model was developed to estimate carbon monoxide exposure. It uses both background and microenvironment carbon monoxide concentrations and derives total microenvironment concentrations by summing the individual exposures. The model uses U.S. Bureau of Census data and a human activity model estimates exposure and dose (Ott 1981; Ott et al. 1988).

PAQM (Personal Air Quality Model) — This model uses hourly sequences of outdoor pollutant concentrations to estimate indoor concentrations. The model uses mass balance equations and compensates for leakage and mechanical ventilation. It also includes population activity (Systems Applications International, Inc., San Rafael, CA).

NEM (National Ambient Air Quality Standards Exposure Model) — This model was developed by the EPA to assess exposure to air pollutants as people move through their normal daily activities. Although the regulatory program is aimed at outdoor pollutants, indoor exposure is accounted for by adding microenvironment-specific concentrations to that portion of the outdoor air that enters the indoor environment (Biller et al. 1981).[3]

Both measurement and models can require detailed, specific data on the physical properties of the subject microenvironment(s) and population exposure and movement into and out of the microenvironment(s). As described by NRC (1991), this is often done through the use of survey research techniques. These surveys typically require the gathering of personal information through questionnaires or personal diaries. To ensure scientific as well as statistical validity, the assessor must carefully consider the selection of the appropriate population and measurement techniques and the development of the questionnaire or diary. Questionnaires are particularly difficult because of the sometimes subjective nature of the information being gathered and the fact that the wording of questions can influence the answers. While there are few widely accepted guidelines for developing questionnaires, research in recent years has explored various aspects of questionnaires and helped to define techniques more likely to be successful (NRC 1991).

---

[3] More recently, the EPA developed a probabilistic version of this model called pNEM (Johnson et al. 1990).

D. Identifying Exposed Populations

Identification of the exposed population can be relatively simple if the object is to evaluate limited indoor microenvironments or a limited number of exposed persons. It can also be difficult if the object is to evaluate a large exposed population or numerous microenvironments. In the simple case, a building investigation might be required and the building occupants might be surveyed using questionnaires or through personal interview. Larger regional investigations, on the other hand, may need to use U.S. Census Bureau data or statistical methods to select residents for questionnaire or interview. Many exposure models incorporate population data obtained from the Census Bureau and can integrate the population data directly with the concentration data to estimate total exposure. As noted by Pandian (1987), the Census Bureau data can be presented in various ways but most often utilizes block group/enumeration district (BG/ED) data at the state and county level. The block group is a contiguous area having an average population of about 1,100 persons and an enumeration district is an area containing about 800 persons; the enumeration district is used when the block group is not defined. The U.S. contains approximately 300,000 individual BG/EDs. Each BG/ED is defined by the population and the centroid (i.e., the center of mass) of its population in that area. The Census Bureau can also provide substantial information on population subgroups, including sex, age, race, economic status, occupational category, and smoking habits.

More difficult investigations may require use of hospital records to identify individuals or proportions of the population that represent sensitive groups. The most widely recognized sensitive groups are shown in Table 5.1. However, many other population groups may be considered. For example, three specific sensitivities have been identified or postulated more recently as potentially resulting from exposure to indoor air pollution.

Table 5.1 Subpopulations with Potentially Increased Responsiveness to Indoor Air Pollutants[a]

Subpopulation	Subpopulation size	Percent of population
Newborns	3,731,000	1.5
Young children	18,128,000	7.5
Elderly	29,172,000	12.1
Heart patients	18,458,000	7.7
Bronchitis sufferers	11,379,000	4.7
Asthma sufferers	9,690,000	4.0
Hay fever sufferers	21,702,000	9.0
Emphysema sufferers	1,998,000	0.8
Smokers	69,852,000	29.9

Source: Indoor Air Report to Congress (EPA 1989).

[a] Statistics from 1986, except smokers in 1983.

Sick Building Syndrome (SBS) — This condition appears to be an adverse sensory reaction of a large portion of the exposed population in a specific building to indoor air that has some unusual characteristic(s). The most prevalent pollutant suspect appears to be volatile organic compounds (VOCs), but it is also believed that many other pollutants and socio-psychological influences (e.g., temperature, noise, illumination, and building location) can influence SBS.

Multiple Chemical Sensitivity (MCS) — This condition may apply to a small population group that is hypersusceptible when exposed to low concentration combinations of multiple chemicals. This syndrome is harder to diagnose and treat than SBS. Validating and evaluating this syndrome are particularly difficult because the pollutants and concentrations to which the individuals are exposed are typically difficult, if not impossible, to measure quantitatively.

Bronchial Asthma — A condition that appears to be growing nationwide for as yet unknown reasons is bronchial asthma. Incidence doubled between 1974 and 1988, roughly paralleling the improvements in indoor air management following the oil embargo. Evidence is increasing that indoor air pollution is a major factor, particularly dust mite allergens. The role of mold and other bioaerosols in the increase in bronchial asthma is less well understood. One recent study (Joseph 1997) notes a correlation between the use of methyl-*tert*-butyl ether (MTBE) in gasoline (used since about 1992 to oxygenate gasoline) and the increase in urban area asthma.

Finally, a unique way to determine when a specific population is exposed to selected indoor air contaminants is to use biological markers. As discussed above, these are cellular, biochemical, or molecular changes that result from specific exposures (NRC 1991). Examples of this use to estimate exposure include: urine cotinine resulting from inhalation of tobacco smoke; carboxyhemoglobin resulting from inhalation of carbon monoxide; and lead in blood, teeth, and hair resulting from inhalation and ingestion of lead.

E. Integrating Exposure Assessment Techniques

In the final analysis, estimating exposure is generally an integrated process using measured, modeled, and gathered data. The assessor must combine information on sources, pollutants, environmental fate and transport, exposed populations, measured or modeled concentrations, and often other information to complete the exposure assessment. Although each of these data sets is important, our understanding of each typically differs with each specific situation. If the exposure link (i.e., source to receptor) is broken by incomplete or incorrect data in any of these areas, the final results can suffer substantially. However, when total (indoor and outdoor) exposure can be estimated with reasonable confidence, it can substantially improve our ability to find more health-protective and cost-effective control strategies to reduce human exposure and ultimately risk. In addition, it can also tell us when current or proposed regulatory or control strategies are misdirected.

The EPA's Indoor Air Report to Congress describes various integrated efforts to estimate exposure through measurement and models. The most widely publicized are the Total Exposure Assessment Methodology (TEAM) studies. These began in the late 1970s. A more complete description is provided in Chapter 8, but the purpose of the TEAM studies was to measure total human exposure to a variety of pollutants, initially focusing on VOCs. Several geographic locations were chosen. Studies typically measured air, water, and exhaled breath samples for a number of target substances. The important early study results included the following:

1. personal air exposure is almost always higher than outdoor exposure;
2. breath levels correlate significantly with personal air exposures but not outdoor air levels;
3. concentrations in the breath identify a number of specific exposures, including smoking, visiting dry cleaners, and pumping gasoline;
4. concentrations in the indoor air demonstrate the presence of smoking and other possible sources, including use of hot water, room air fresheners, toilet bowl deodorizers, and moth crystals; and
5. for all chemicals except halomethanes, inhalation provides greater than 99% of the exposure.

The integration of exposure techniques is best demonstrated by describing current activities on several air pollutants that are both outdoor and indoor air pollutants. A more complete discussion of these pollutants can be found in NRC 1991.

1. VOCs

There are hundreds, if not thousands, of volatile organic compounds (VOCs) to which people can be exposed each day. Tens of thousands are in commerce, with new compounds and new uses being invented each year. Many of these VOCs can be dangerous, some at virtually any concentration but most only at concentrations well above those typically observed indoors. These VOCs can arise in the indoors (e.g., from painting and cleaning), and be brought into the indoors (e.g., with drinking water and dry cleaning), and VOCs in outdoor air can infiltrate indoors. VOCs are present indoors at concentrations believed to be well below levels of potential concern in the workplace. However, some researchers today postulate possible adverse effects in sensitive persons to the combined exposure to numerous, low-concentration VOCs. The dominant exposure pathway for VOCs is inhalation, and conditions are generally not favorable indoors for significant chemical or physical change through processes such as chemical reaction or degradation.

VOCs are typically identified and quantified using gas chromatography and mass spectrometry, analytical processes that are uniformly complex and expensive. Although some individual VOCs can be monitored in real-time, a full analysis usually requires that samples be collected and returned for analysis at a laboratory. In many instances, the VOCs are captured on a sorbent before analysis, and the most widely used sorbents are subject to various limitations. Newer canister methods appear to be capable of providing improved accuracy.

The EPA regulates VOCs as air pollutants under the Clean Air Act (CAA). First, VOCs are regulated under sections 108 to 111 as precursors to ozone, a listed criteria air pollutant. In general, a national ambient air quality standard (NAAQS) is established for criteria pollutants, and sources of the pollutants or precursors are regulated using a combined federal and state program. Estimated exposure to the criteria pollutant is an important factor in determining the potential for adverse health effects and, thus, the level and coverage of the NAAQS. Once the NAAQS is established, geographic regions that consistently exceed a NAAQS are said to be in nonattainment. Nonattainment areas must conduct studies, often including regional modeling, to determine the best combination of controls and emission reductions that will attain and maintain for the long term the ambient concentrations at levels below the NAAQS. Source emissions primarily determine regulatory applicability, coverage, and degree. VOCs are known to react photochemically with nitrogen oxides (NO_x) to form tropospheric ozone, thus, VOCs are a precursor to ozone. In the regulation of ozone through the years, the primary control focus has been VOCs because the link between VOC and ozone reduction is relatively well understood while the link between NO_x and ozone reduction is much more complex. As ozone nonattainment persists in many areas of the country, and the VOCs available for further reduction diminish, legislative and regulatory attention on NO_x has recently increased.

Second, specific VOCs are also regulated as hazardous air pollutants under section 112. Vinyl chloride and benzene are VOCs specifically regulated as HAPs under the 1970 Amendments to the CAA; over 100 other VOCs are listed as HAPs in the 1990 Amendments to the CAA. The original vinyl chloride and benzene regulations were promulgated before exposure and risk assessment guidelines were fully developed by the EPA, and both regulations entered protracted litigation. At the heart of the litigation was the EPA's belief that cost and technical feasibility should be a part of the HAP decision process and the belief of environmental groups that they should not. As discussed in Chapter 1, the U.S. Court of Appeals ruled against the EPA in 1986 on vinyl chloride and directed that the Agency establish standards for HAPs primarily using risk as the guide. Regulations for benzene were promulgated using the new guidance, but soon thereafter the 1990 Amendments to the CAA were enacted and changed the regulatory process for HAPs. Section 112 of the 1990 Amendments first requires application of maximum achievable control technology (MACT) on all major sources of the listed HAPs. Eight years later, the risks remaining after MACT are to be evaluated and additional standards promulgated if needed to protect the public health with an ample margin of safety. In other words, the initial phase requires only installation of available control technology while the second phase requires consideration of population exposures and risks.

These two CAA programs, however, present several problems. For example, the requirements for VOCs focus only on outdoor sources and, until recently, rested on the assumption that people are exposed continuously to measured or modeled outdoor concentrations of the air pollutants of concern. However, studies begun in the 1970s (e.g., the TEAM studies discussed above) determined that this is generally not the case and that a significant, and often predominant, portion of the average person's time is spent indoors. In addition, exposure to indoor sources of air pollutants were found to influence substantially the amounts and types of VOCs to which people are exposed. Thus, because of the legislative requirements of the CAA we find that some current HAP regulations do not address the predominant exposure pathways of the HAPs. For example, several regulations are in place that reduce industrial (e.g., petroleum refineries) and commercial (e.g., gasoline refueling) emissions of benzene, but research has shown that only a very small fraction of the population exposure to benzene results from industrial and commercial emissions. Much more important indoor exposure sources of benzene include smoking, gasoline fumes from pumping gasoline and from attached garages, products in the home, and painting. Clearly, better monitoring and modeling efforts are required to estimate more accurately indoor VOC exposures and ultimately risks. Improved methods are beginning to be used (Johnson et al. 1990) but many remain to be developed and verified for the majority of VOCs to be regulated.

2. Polycyclic Aromatic Hydrocarbons (PAHs)

PAHs are produced largely from the incomplete combustion of organic materials. The number of PAHs is large and several specific PAHs are known to cause cancer in animals or humans (e.g., testicular cancer in chimney sweeps in 19th century England is largely ascribed to PAHs in the soot). However, PAH concentrations in the ambient environment generally are low and difficult to measure; again, samples must be collected and taken to the laboratory for analysis. These difficulties led many regulators to focus on one PAH, benzo(a)pyrene (BaP), a known animal carcinogen and a PAH that has been measured widely. For simplicity, BaP is used as an indicator substance of the presence and potency of PAH mixtures; however, many investigators believe that use of BaP is not an effective or accurate means of estimating exposures and risks of all PAHs.

Because of their high molecular weights, PAHs do not generally exist as vapors but are adsorbed onto the surface of particulate matter. This results in potential exposures through multiple pathways. For example, particulate matter with a diameter of 10 microns or less can be inhaled; larger particles can deposit indoors and outdoors where intake can occur through skin contact; and larger particles can deposit on plants and water outdoors that can be consumed by animals or brought to humans for consumption. Environmental fate is generally not significant because conditions are not favorable indoors for significant chemical or physical change through processes such as reaction or degradation.

There is little evidence linking community concentrations of PAHs to cancer, although a major reason for this lack is that the small risk of cancer predicted from outdoor exposure to PAHs is completely overshadowed by the large risk of cancer estimated to result from exposure to tobacco smoke, which also contains PAHs. Studies similar to the TEAM studies (Lioy et al. 1988) also show that a large fraction of the average person's exposure to PAHs results from indoor rather than outdoor sources; important indoor sources of PAHs include fuel combustion, food preparation, smoking, and unvented space heaters. However, these studies also show that indoor exposures to PAHs more closely track outdoor exposures in the homes of nonsmokers. As noted above, PAHs that are released from outdoor sources as particles or adsorbed onto particles gravimetrically deposit onto crops or other surfaces for ultimate absorption and human consumption. Studies have found low concentrations of PAHs in many foods, even foods grown far away from major industrial sources. Again, better monitoring and modeling efforts are required to estimate more accurately PAH exposures and risks.

3. Lead

Lead has long been known to be detrimental to human health. However, it was not widely regulated until the 1970 CAA and the EPA's subsequent listing of lead and its compounds (measured as elemental lead) as a criteria air pollutant under section 108, and the listing of lead compounds (although not elemental lead) as a HAP under section 112 of the 1990 Amendments to the CAA. As with PAHs, potential exposures to lead are multipathway. Lead in the environment exists in a solid form with small particles being capable of being inhaled and larger particles depositing for subsequent intake. Lead is also relatively easy to measure.

Initial regulatory concerns with lead began with the growing evidence of increased blood lead levels in children exposed to lead in soil near heavily traveled motor vehicle corridors and near large sources of lead emissions, and the knowledge of the adverse health effects that can result from exposure to lead. Lead compounds were used for many years as octane enhancers and engine part lubricants until they began to be phased out with the introduction of catalytic converters in the early 1970s (lead destroys the catalyst activity). Lead compounds were also widely used in paints, and many primary and secondary smelters emitted large quantities of lead for deposition onto the earth's surface. These latter sources often resulted in high quantities of lead in the soil and, ultimately, in humans in high traffic areas, in older homes, and near industrial facilities using or emitting lead. The concerted efforts to reduce lead exposures through the introduction of lead-free gasoline, restrictions on paint composition and coverage, and regulations on industrial sources of lead resulted in significant measured reductions in lead concentrations in humans and particularly children. However, recent studies and continuing studies conducted and reported upon by the Agency for Toxic Substances and Disease Registry (ATSDR) still show clinically significant elevations of lead in the blood of many children (ATSDR 1988). Children are particularly at risk to adverse effects resulting from lead. These concerns led to continuing exposure assessments that include both ambient monitoring and measurement, and human testing. In the late 1980s, the EPA maintained 353 monitoring stations in the U.S. to directly measure lead in the environment.

4. Environmental Tobacco Smoke

As described by the NRC (1986 and 1991), while not regulated as an air pollutant, the health hazards of smoking have been studied for many years and are well recognized. Environmental tobacco smoke (ETS) is that portion of the smoke to which nonsmokers are exposed and includes sidestream smoke (emitted from the burning end of the product during puffs) and mainstream smoke (smoke that is drawn into the mouth and then exhaled). Sidestream smoke is the major source of ETS.

ETS is rarely measured directly because it is a complex mixture of substances. More than 3,800 compounds in particle and vapor phases have been identified in cigarette smoke, with major components including nicotine, carbon monoxide, particulate matter, aromatic hydrocarbons, and numerous tobacco-specific chemicals. Epidemiological studies are hampered by the fact that exposures to ETS, as with many other indoor air pollutants, occur at wide ranges of concentrations, over highly variable periods, and in a wide variety of indoor and outdoor environments. Personal monitoring can determine a person's total exposure to ETS, but a questionnaire or diary is required to determine how the person is exposed. Questionnaires are frequently shown to be biased. For example, urine testing that measures metabolites of nicotine has indicated that some smokers will claim to be nonsmokers for reasons including the social stigma attached to smoking. In this instance, a person's total exposure might be indicated to be the result of residential or workplace exposure based on the questionnaire, but could actually result from personal smoking that is not admitted.

Because of the large number of measurable compounds in ETS, researchers often focus on target compounds. Two important markers for exposure studies are vapor-phase nicotine and respirable suspended particles (RSP). Almost all (95% or more) of nicotine in ETS appears to be in the vapor phase and tobacco is the predominant source of nicotine. Tobacco burning also emits large amounts of RSP. Both are relatively easily measured, although a number of assumptions are usually necessary in order to estimate exposure. As noted above, cotinine is widely used as a biological marker indicating exposure to nicotine. Other biological markers are also used, including thiocyanate, carboxyhemoglobin, aromatic amines, and protein and DNA adducts. Again, these biological markers indicate that exposure has occurred but cannot precisely pinpoint the source; they can also vary across the population. Some markers are not specific; for example, carboxyhemoglobin also results from exposure to carbon monoxide from sources other than cigarettes.

F. Uncertainty in Exposure Characterization

As described by Patrick (1992), uncertainty in exposure assessment can arise from: (1) variations in methods or models used, (2) variations in inputs to those methods and models, (3) imprecise knowledge of the underlying science, and (4) natural variability. Although Chapter 7 provides a more detailed discussion of uncertainty in risk assessment, several uncertainty issues specific to exposure assessment are summarized here and ranges of potential uncertainty are estimated.

Exposure begins with the emission of pollutants into the air from a source. These emissions can be particles (of varying sizes) or vapors and can be released from a wide variety of sources and under widely varying conditions, with height, temperature, and velocity among the more important. Emissions from a typical air pollutant source vary both temporally and spatially. The emissions are dispersed in all directions away from the source(s). This potential for three-dimensional dispersion is a unique feature of air pollutants when compared to water and soil pollutants, where the releases typically are much more channeled or directed (e.g., into the surface water, ground water, or subsurface soil). Dispersion into the air is influenced both by meteorology and topography. Meteorology moves and influences the emissions and topography can channel or contain emissions in specific directions or locations. Generally, emission concentrations reduce in all directions as the emissions fill an ever greater volume of air. Many emissions can also degrade (e.g., under the action of sunlight) or react (e.g., through oxidation) to form other pollutants, of more or less concern. The population closest to the source is generally exposed to the highest concentrations of the emissions, and populations farther away can be affected more by reaction products or products of degradation. Importantly, population is not a single affected entity. Rather, it can be an array of individuals who respond in many different ways depending upon their characteristics (e.g., age, sex, health, and life-style). Finally, assessments can be made using gathered data or modeled assumptions. Models are frequently used because sampling and analysis costs can be high, the emissions may have ceased or changed, large areas may be involved, and other factors. Both model variations and model input variations can occur.

This summary description of uncertainty in exposure focuses on six important areas of analysis where there are uncertainties. These areas were described and quantified more completely by the author (Patrick 1992) and are summarized here.

1. Location of Exposed Population

Uncertainties in the location of the exposed population can result in overestimates or underestimates in exposure and risk, particularly for those located near the sources of emissions. This uncertainty arises because people are uniquely distributed around every source of air pollutants, and there are so many people (air pollution can disperse many miles) and sources that a complete knowledge of their location is practically impossible. Complicating this further is the fact that these same people are also uniquely influenced by all of the other sources (i.e., stationary and mobile, indoor and outdoor) of air pollutants to which they are exposed. Early exposure and risk assessments tended to ignore the population distribution, focusing on the maximum concentration and assuming that someone could be exposed to that concentration. This was done because it was both easier and cheaper to use the maximum than to include all potentially exposed people. In addition, the emerging fields of exposure and risk assessment were associated with sufficient overall uncertainty to argue logically for a worst-case estimate in order to be health-protective. As the degree of sophistication in exposure assessment improved, broader populations were considered. As discussed earlier, Bureau of Census data are provided on a block group/enumeration district (BG/ED) basis. For each BG/ED, a total population is provided along with the longitude and latitude of the centroid of the population. This allows easy computer manipulation. Since exposure at actual residences is not available and not reasonably obtainable, population in a BG/ED is generally assumed to reside at the centroid of the BG/ED. This can result in underestimates or overestimates of exposure and risk because some people in the BG/ED live closer to the source and some live farther away from the source. Maximum exposure and risk are generally underestimated, particularly close to the source of emissions because the centroid is usually not at the point of maximum concentration. Average exposure and risk may be overestimated or underestimated. Early exposure assessments also inaccurately assumed that people resided at the centroid for their lifetime (70 years), indoor and outdoor air exposures were the same, emissions remained constant for 70 years, and there was no population movement or growth. While more recent exposure and risk assessments may still use Bureau of Census data, many of the accompanying assumptions are improved significantly.

2. Population Lifestyles and Activity Patterns

Uncertainties in the knowledge of population lifestyles and activity patterns generally result in overestimates in exposure and risk. As noted above, early exposure and risk assessments typically assumed that the exposed population resided at the same location for 70 years and that indoor and outdoor exposures were the same. This obviously ignores the fact that people move, work, and play, and the present knowledge that indoor exposures are typically substantially different from outdoor exposures. The EPA and others have attempted to describe population activity patterns more precisely (EPA 1995). While these patterns vary widely, just as people do, studies and surveys have developed statistical distributions (e.g., with averages and standard deviations) for many of the important variables. Often, then, values are chosen based on the average and two standard deviations to represent 95% of the population of the variables. This is a more realistic means of describing the population than assuming that all of the population is the same as the maximum or worst case. One difficulty in applying these more realistic conditions, however, is that environmental laws and the regulations that result from them usually are written for a single medium (e.g., air, water, solid, and hazardous waste) and do not easily accommodate other considerations. For example, indoor exposures and risks are only recently being considered as part of the outdoor air exposure and risk assessments conducted under the CAA, even though researchers and regulators have known for years that indoor and outdoor air concentrations can be significantly different and that people spend the vast majority of their time indoors. The EPA's Draft Exposure Factors Handbook (EPA 1995) also provides a wealth of information on population lifestyles and activity patterns as well as human intake distributions discussed below.

3. Human Intake

Uncertainties in human intake can result in overestimates or underestimates of exposure and risk. Again, because of the single-media focus of environmental laws and regulations, early exposure and risk assessments conducted for air pollutants generally included only human intake by inhalation. This procedure can lead to inaccuracies in several ways. For example, many air pollutants (e.g., particulates) can be inhaled but they can also deposit onto the ground, plants, and surface waters and be taken up by plants and other organisms. These pollutants can affect humans who eat the contaminated plants or fish, drink the contaminated water, or come into contact with pollutants that can be absorbed or ingested; they can also independently affect those plants and animals. In addition, other pollutants may be present in these materials that can add to the effects. Furthermore, pollutants taken into the body were assumed to be completely absorbed and available to cause adverse effects. Researchers now know that pollutants are biologically effective in many different ways depending upon the chemical and physical characteristics of the pollutant and the physical and biological processes that take place in the human body. Intake also can depend upon such characteristics as age, sex, weight, and breathing rate, which differ widely. Early exposure and risk assessments rarely took into account the characteristics of special groups such as children, the elderly, and the infirm, and assumed that all exposure was by adults at one body weight of 70 kg.

4. Emission Characteristics

Uncertainties in emission characterization can result in overestimates or underestimates of exposure and risk. These uncertainties arise for many reasons, not the least of which is the technical difficulty in determining on a continuous basis the true emissions from a source. Only a few sources of air pollutants in the U.S. and the world have monitors that can provide continuous real-time emissions data. This results because of the high cost of continuous monitoring and the fact that these monitors have been developed for only a few air pollutants. In most cases, air pollutant regulations in the past required annual testing using a stack test (i.e., usually three separate tests with the results averaged) during a period of normal source operation. Depending upon the inherent variability of the source, this may overestimate or underestimate true emissions.

This procedure of annual testing of air pollutants may be changing in the U.S. In 1997, two significant actions were taken by the EPA. First, based on a court ruling, the EPA published regulations providing that "any credible evidence" be used to assess compliance and that regulators could issue violations and require fines to be paid based on the credible evidence (62, FR 8314, February 24, 1997). While the decision was specific to excess opacity as an indicator of excess particulate emissions, it appears that almost any variable that is monitored frequently (e.g., temperature, pressure, or nonpollutant emissions) could be used to determine compliance if a reasonable relationship could be established between the variable and the air pollutant emissions. U.S. industry vigorously contested this ruling and by mid-1997 nearly 100 lawsuits had been filed seeking its overturn. Second, the 1990 Amendments to the CAA required continuous compliance with air pollution regulations. The EPA responded to this requirement by proposing enhanced monitoring regulations that were immediately challenged by industry as being onerous and excessively costly. The EPA finalized a scaled back Compliance Assurance Monitoring (CAM) rule that was intended to address industry concerns (62 FR 54900, October 22, 1997). However, there continue to be significant disagreements with the plans and, at the time of this writing, it appeared likely that this rule also would be litigated. Depending upon the outcome of the litigation, emissions monitoring in the future may include other related variables and it may be required much more frequently than once per year.

5. Duration and Frequency of Exposure

Uncertainties in the duration and frequency of exposure usually result in overestimates in exposure and risk. In early exposure and risk assessments, continuous exposure at the maximum possible ambient concentrations outside the fenceline of the emission source was often assumed. This hindered the need for more detailed analysis but it almost always resulted in significant overestimates of exposure and risk. Dispersion modeling is capable of providing much more accurate estimates of long-term average and short-term peak concentrations of air pollutants resulting from a source of air pollutant emissions. These models, originally developed for use on mainframe computers, have been improved over the past three decades and now are available in forms that have been generally validated to about 50 kilometers from typical sources. Models now can account for wide variations in source configuration, weather, and topography, and often can be used on personal computers. The initial important step in the process, however, is accurate knowledge of the emissions and their characteristics from the source. To the extent that these can be determined with precision, the dispersion model can then provide reasonable estimates of the average and peak impacts on the surrounding community under certain assumed conditions. While a dispersion model is unable to provide estimates for unknown future emissions, it can provide estimates for ranges of expected conditions.

6. Environmental Fate and Transport

Uncertainties in the knowledge of environmental fate and transport of outdoor air pollutants can result in overestimates or underestimates of exposure and risk. Outdoor air pollutants can be influenced in many ways by fate and transport. An important example of environmental fate is the photochemical conversion of nitrogen oxides and volatile organic compounds into tropospheric ozone; an important example of environmental transport is the movement of sulfur oxides and nitrogen oxides many hundreds of miles from Midwestern power plants with deposition of acidic materials in northeastern U.S. and Canada. These processes are much less important, although not totally absent, in the assessment of indoor air pollutants. For example, outdoor air pollutants may be transported long distances and have undergone some chemical alteration before being taken into an indoor environment where exposure can occur. Some other pollutants generated or brought into the indoor environment may also change form depending upon the other pollutants (i.e., coreactants), materials (i.e., reactive surfaces), or processes (i.e., combustion sources) that are present.

BIBLIOGRAPHY

Agency for Toxic Substances and Disease Registry (ATSDR). 1988. *The Nature and Extent of Lead Poisoning in Children in the United States: A Report to Congress, July 1988*, U.S. Department of Health and Human Services, Public Health Service: Atlanta, GA.

Biller, W.F., Feagans, T., et al. 1981. A General Model for Estimating Exposure Associated with Alternative NAAQS, Paper No. 81-18.4, Presented at the 74th Annual Meeting of the Air Pollution Control Association, Phildelphia, PA, June, 1981.

Billick, I.H. 1990. Estimates of population exposure to nitrogen dioxide, *Toxicology and Industrial Health* 6(2):325.

Brauer, M., Bartlett, K., et al. 1996. Assessment of particulate concentrations from domestic biomass combustion in rural Mexico, *Environmental Science and Technology* 30(1):104–109.

Chemical Manufacturers Association. 1991a. *Analysis of the Impact of Exposure Assumptions on Risk Assessment of Chemicals in the Environment. Phase I: Evaluation of Existing Exposure Assessment Assumptions*, prepared by VERSAR: Springfield, VA.

Chemical Manufacturers Association. 1991b. *Analysis of the Impact of Exposure Assumptions on Risk Assessment of Chemicals in the Environment. Phase II: Uncertainty Analyses of Existing Exposure Assessment Methods*, prepared by VERSAR: Springfield, VA.

Chemical Manufacturers Association. 1991c. *Analysis of the Impact of Exposure Assumptions on Risk Assessment of Chemicals in the Environment. Phase III: Evaluation and Recommendation of Alternative Approaches*, prepared by VERSAR, Springfield: VA.

De Koster, J.A., Thorne, P.S. 1995. Bioaerosol concentrations in noncompliant, compliant, and intervention homes in the midwest, *American Industrial Hygiene Association Journal* 56:573–580.

Duan, N. 1985. *Application of the Microenvironment Monitoring Approach to Assess Human Exposure to Carbon Monoxide*, R-3222-EPA, prepared for the U.S. Environmental Protection Agency by Rand, Santa Monica, CA.

Ekberg, L.E. 1995. Concentrations of NO_2 and other traffic related contaminants in office buildings located in urban environments, *Building and Environment* 30(2):293–298.

Environmental Protection Agency. 1979. Methodology for Estimating Direct Exposure to New Chemical Substances, Report No. EPA 560/13-79-008, Office of Toxic Substances, U.S. Environmental Protection Agency: Washington, DC.

Environmental Protection Agency (EPA). 1987a. *Compendium of Methods for the Determination of Air Pollutants in Indoor Air*, Office of Research and Development, Environmental Monitoring Systems Laboratory, U.S Environmental Protection Agency: Washington, DC.

Environmental Protection Agency (EPA). 1987b. *Evaluation of Existing Total Human Exposure Models*, CR 812189-01-0, Office of Research and Development, Environmental Monitoring Systems Laboratory, U.S. Environmental Protection Agency.

Environmental Protection Agency (EPA). 1989. Report to Congress on Indoor Air Quality, Report No. FPA/400/1-89/001A, Office of Air and Radiation and Office of Research and Development, U.S Environmental Protection Agency: Washington, DC, August 1989.

Environmental Protection Agency (EPA). 1992a. A Tiered Modeling Approach for Assessing the Risks Due to Sources of Hazardous Air Pollutants, Report No. EPA-450/4-92-001, Office of Air Quality Planning and Standards, Technical Support Division, U.S. Environmental Protection Agency.

Environmental Protection Agency (EPA). 1992b. Guidelines for Exposure Assessment, *57 FR 22887*, May 29, 1992.

Environmental Protection Agency (EPA). 1995. *Exposure Factors Handbook, Draft,* Report No. EPA/600/P-95/0002A, Office of Research and Development, U.S. Environmental Protection Agency.

Fischer, M.L., Bentley, A.J., et al. 1996. Factors affecting indoor air concentrations of volatile organic compounds at a site of subsurface gasoline contamination, *Environmental Science and Technology* 30(10):2948.

Foster, S.A., Chrostowski, P.C. 1986. Integrated Household Exposure Model for use of Tap Water Contaminated with Volatile Organic Chemicals, for presentation at the 79th annual meeting of APCA, Minneapolis, MN, June 22–27, 1986.

Foster, S.A., Chrostowski, P.C. 1987. Inhalation Exposures to Volatile Organic Contaminants in the Shower, for presentation at the 80th annual meeting of APCA, New York, NY, June 21–26, 1987.

Giardino, N.J., Hageman, J.P. 1996. Pilot study of radon volatilization from showers with implications for dose, *Environmental Science and Technology* 30(4):1242–1244.

Jarvis, D., Chinn, S., et al. 1996. Association of respiratory symptoms and lung function in young adults with use of domestic gas appliances, *Lancet* 347:426–431.

Jayjock, M.A., Hawkins, N.C. 1993. A proposal for improving the role of exposure modeling in risk assessment, *American Industrial Hygiene Association Journal* 54(12):733–741.

Jenkins, P.L. et al. 1992. Activity patterns of Californians: Use of and proximity to indoor pollutant sources, *Atmospheric Environment* 26A(12):2141.

Jenkins, R.A., Palausky, A., et al. 1996. Exposure to environmental tobacco smoke in 16 cities in the United States as determined by personal breathing zone air sampling, *Journal of Exposure Analysis and Environmental Epidemiology* 6(4):473–502.

Johnson, T.R., Paul, R., et al. 1990. Estimation of Ozone Exposure in Houston Using a Probabilistic Version of NEM, Paper No. 90-150.1, for presentation at the 83rd Annual Meeting of the Air and Waste Management Association, Pittsburgh, PA, June 1990.

Joseph, P.M. 1997. Changes in Disease Rates in Philadelphia Following Introduction of Oxygenated Gasoline, Paper No. 97-TA34.02, for presentation at the 90th Annual Meeting of the Air and Waste Management Association, Toronto, ON, June 1997.

Kaplan, M.P., Brandt-Rauf, P., et al. 1993. Residential releases of number 2 fuel oil: A contributor to indoor air pollution, *American Journal of Public Health* 83(1):84–88.

Klepeis, N.E., Ott, W.R., et al. 1996. A multiple-smoker model for predicting indoor air quality in public lounges, *Environmental Science and Technology* 30(9):2813–2820.

Kostiainen, R. 1995. Volatile organic compounds in the indoor air of normal and sick houses, *Atmospheric Environment* 29(6):693–702.

Lioy, P.J., Greenberg, A. 1990. Factors associated with human exposures to polycyclic aromatic hydrocarbons, *Toxicology and Industrial Health* 6(2):209.

Lioy, P.J., Waldman, J., et al. 1988. The total human environmental exposure study (THEES) to benzo(a)pyrene: Comparison of the inhalation and food pathways, *Archives of Environmental Health* 43:304–312.

Lippmann, M. 1990. Man-made mineral fibers (MMMF): Human exposures and health risk assessment, *Toxicology and Industrial Health* 6(2):225.

Mahanama, K.R.R., Daisey, J.M. 1996. Volatile *n*-nitrosamines in environmental tobacco smoke: Sampling, analysis, emission factors, and indoor air exposures, *Environmental Science and Technology* 30(5):1477–1484.

McAughey, J.J., Pritchard, J.N., et al. 1990. Risk assessment of exposure to indoor air pollutants, *Environmental Technology* 11:295.

National Research Council. 1986. *Environmental Tobacco Smoke: Measuring Exposures and Assessing Health Effects,* prepared by the Committee on Passive Smoking, Board of Environmental Studies and Toxicology, National Academy Press: Washington, DC.

National Research Council. 1991. Human Exposure Assessment for Airborne Pollutants; Advances and Opportunities, National Academy of Sciences: Washington, D.C.

Ott, W.R. 1981. Computer Simulation of Human Air Pollution Exposures to Carbon Monoxide, Paper No. 81-57.6, for presentation at the 74th Annual Meeting of the Air Pollution Control Association, Philadelphia, PA, June 1981.

Ott, W.R., Thomas, J., et al. 1988. Validation of the simulation of human activity and pollution exposure (SHAPE) model using paired days from the Denver, CO, carbon monoxide field study, *Atmospheric Environment* 22:2101–2113.

Pandey, J.S., Mude, S., et al. 1994. Comparing indoor air pollution health risks in India and the United States, *Journal of Environmental Systems* 23(2):179–194.

Pandian, M.D. 1987. Evaluation of Existing Total Human Exposure Models, Report No. CR 812189-01-0, U.S. Environmental Protection Agency, Office of Research and Development, Environmental Monitoring Systems Laboratory: Washington, DC.

Patrick, D.R. 1992. The Impact of Exposure Assessment Assumptions and Procedures on Estimates of Risk Associated with Exposure to Toxic Air Pollutants, Paper No. 92-95.02, presented at the 85th Annual Meeting of the Air and Waste Management Association, Kansas City, MO, June 1992.

Platts-Mills, T.A.E., Chapman, M.D., et al. 1990. Establishing health standards for indoor foreign proteins related to asthma: Dust mite, cat and cockroach, *Toxicology and Industrial Health* 6(2):197.

Repace, J.L., Lowrey, A.H. 1993. An enforceable indoor air quality standard for environmental tobacco smoke in the workplace, *Risk Analysis* 13(4):463–475.

Samet, J.M. 1995. Asthma and the environment: Do environmental factors affect the incidence and prognosis of asthma?, *Toxicology Letters* 82/83:33–38.

Sidhu, K.S., Hesse, J.L., et al. 1993. Indoor air: Potential health risks related to residential wood smoke, as determined under the assumptions of the U.S. EPA risk assessment model, *Indoor Environment* 1993(2):92–97.

Sram, R.J., Benes, I., et al. 1996. Teplice program—the impact of air pollution on human health, *Environmental Health Perspectives*, 104(4):699–714.

State of California. 1989. *Fundamentals of Exposure Assessment, Toxic Substances Control Division*, Department of Health Services, prepared by Clement Associates for Canonie Engineers, Inc., February, 1989.

Stevens, R.K., Lewis, C.W., et al. 1990. Sources of mutagenic activity in urban fine particles, *Toxicology and Industrial Health* 6(5):81–94.

Stolwijk, J.A.J. 1990. Assessment of population exposure and carcinogenic risk posed by volatile organic compounds in indoor air, *Risk Analysis* 10(1):49.

Stolwijk, J.A.J. 1992. Risk assessment of acute health and comfort effects of indoor air pollution, *Annals of the New York Academy of Sciences* 641:56–62.

Wallace, J.C., Basu, I., et al. 1996. Sampling and analysis artifacts caused by elevated indoor air polychlorinated biphenyl concentrations, *Environmental Science and Technology* 30(9):2730–2734.

Wallace, J.C., Brzuzy, L.P., et al. 1996. Case study of organochlorine pesticides in the indoor air of a home, *Environmental Science and Technology* 30(9):2715–2718.

Weschler, C.J., Shields, H.C. 1996. Production of the hydroxyl radical in indoor air, *Environmental Science and Technology* 30(11):3250.

Zweidinger, R.B., Stevens, R.K., et al. 1990. Identification of volatile hydrocarbons as mobile source tracers for fine-particulate organics, *Environmental Science and Technology* 24:538–542.

CHAPTER **6**

Risk Characterization

Roy E. Albert

CONTENTS

I. INTRODUCTION

In the 15th century, Theophrastus Bombastus Hohenheim (Paracelsus) announced that everything is toxic; it is just a matter of dose. This is the one thing in toxicology that almost everyone agrees with. It follows, therefore, that every agent, within reason, ought to have some form of control whether by a recommendation on intake limits or an enforceable regulatory exposure standard.

Risk assessment provides the basis for deciding how, and to what extent, a given agent (e.g., a carcinogen or noncarcinogen) should be regulated and, if so, in what media, with what toxicological endpoint, and to what degree. Risk assessment has become a powerful tool because it provides a systematic way of organizing what is known and not known about the toxicology of an agent and the interpretation(s) of the data as the basis for making regulatory decisions. The limitations of risk assessments, if competently performed, are not a function of the process itself but a reflection on the limitations of existing knowledge, whether specific to the agent or to the understanding of basic mechanisms that relate to the particular agent. Even

1-56670-323-9/99/$0.00+$.50
© 1999 by CRC Press LLC

though risk assessment began in a formalized way in the area of carcinogenesis, the process is applicable to all forms of toxicity.

Risk assessment began inadvertently. In the mid-1970s, the EPA was heavily criticized by the scientific community and industry because of the attempt by its lawyers to reach general agreement on a rigid set of criteria for carcinogenic properties, called cancer principles, in order to shorten the legal hearing process (Albert 1994). The EPA decided, as a response to this criticism, on a policy that called for balancing risks and benefits as the basis for regulation. This, in turn, required guidelines on how to go about evaluating health risks of suspected carcinogens. The guidelines divided the assessment process into a qualitative (hazard) assessment and a quantitative (dose–response–exposure) assessment. Both components required a variety of disciplines:

- chemistry for the basic properties and modes of interaction;
- detoxification processes;
- biochemical defense mechanisms;
- pharmacokinetic behavior according to the route of exposure;
- genetics for the genotoxic interactions with somatic and germ cells;
- experimental pathology for the outcomes of animal bioassay;
- epidemiology for human studies;
- engineering for characterization of environmental transport and exposure; and
- biostatistics for evaluation of all of the component parts of the assessment and particularly the dose–response relationships.

After presentation of each individual component of the risk assessment, it is necessary to put the outcomes together to make a coherent statement about the two essential questions posed by a risk assessment.

1. How likely is the agent to be a human carcinogen or other form of toxicant?
2. How much cancer or other forms of toxicity will the agent produce given the existing exposure scenarios?

In seeking to answer these questions, the mental processes are similar to those used to make any decision and are weighted according to their relative importance, and the alternative possibilities are considered according to these weighted factors. With carcinogenesis, the rank order of importance of evidence is relatively noncontroversial. There is primary evidence, namely of cancer induction, most importantly in humans although infrequently available. There is also evidence in animals where the greater the range of species that respond, the greater the weight of evidence. Next, there are secondary lines of evidence, such as the chemistry of the agent, which can stand alone or modify the primary evidence. For example, the analyst might explore whether a substance is electrophilic (meaning adduct-forming on macromolecules such as DNA), whether it can be metabolically activated to an electrophilic form, and whether it is mutagenic in tests systems including bacteria, yeasts, and mammals. In addition, how do its pharmacokinetics (i.e., absorption, chemical reaction rates, enzymatic reaction rates) and absorption characteristics affect its ability to attack different organs by different routes of exposure? The impact of each of these factors is necessarily modulated by the quality and scope of the data and the nature of the elicited responses. Essentially the same considerations apply to most toxicants whether carcinogens or noncarcinogens. These modifiers can make the risk assessments of individual agents highly controversial. It is useful to capsulate each of these risk assessment components according to a level of evidence, such as that used by the International Agency for Research on Cancer (IARC) for carcinogens (e.g., sufficient or limited) (IARC 1987). This permits the assemblage of the component parts of risk assessment into composite weight-of-evidence categories such as definite, probable, or possible carcinogens. These categories can be used as priorities for regulatory action or in deciding whether to regulate an agent on the basis of its carcinogenicity or on some other form of toxicity. All carcinogens are notably toxic aside from their carcinogenic properties.

According to the National Research Council (NRC) documents on risk assessment (NRC 1994; NRC 1983), risk characterization is the combining of dose–response modeling and exposure assessment to yield numerical estimates of risk. By contrast, the EPA in its guidelines (EPA 1976; EPA 1986; EPA 1996a) defines risk characterization more broadly. It includes the quantitative aspects of risk characterization and an overview of the complete health risk assessment to include the qualitative or hazard assessment. The EPA justified its position on the grounds that all evaluations of risk involve a two-step process: (1) how likely is the risk to occur? and (2) what are the consequences if it does occur (Albert et al. 1977)? For example, the risk of a child falling is very high, but the consequences are generally small, whereas the risk of a nuclear power reactor accidentally releasing massive quantities of fission products into the environment is small but the consequences are many. This two-step evaluation of risk has its analogy in carcinogen risk assessment, in terms of qualitative and quantitative assessment, as indicated above. A risk assessment that does not include both aspects is incomplete.

The idea that all carcinogens are alike is also incorrect. The EPA explicitly adopted a weight-of-evidence approach, generally eschewing flat declarations of whether the agent is or is not a carcinogen, because the issue is whether the agent is a *human* carcinogen. The determination of that property is a complex matter and only in a limited number of instances can one say with certainty that a substance is definitely a human carcinogen. IARC recognized the same principle and summarized its weight-of-evidence judgments in a descriptive numerical code (IARC 1987), which the EPA essentially adopted.

Confusion arises because the term *risk* has two meanings: (1) it means the quantitative nature of the toxic damage as used by the NRC, and (2) it is used at the same time in an overarching sense to indicate both the qualitative (hazard) and quantitative (dose–response and exposure assessment) components of the health assessment. The term *risk assessment* refers to the entire field in all its aspects. It might be less confusing to have the "Risk Characterization" section restricted to the quantitative aspects of risk as described by the NRC and have a separate section, possibly called "Health Assessment Summary," to pull together the entire risk assessment. This function is assigned in the EPA guidelines to a subsection of Risk Characterization, called "Summary of Risk Characterization."

There can be different objectives to risk assessments. For example, one is for regulatory agencies to decide whether regulation, both in kind and degree, is appropriate for toxicants already in use or projected for use; another is for the producers of products who must make decisions to continue the process of bringing a new commodity to the market at all or in modified form. Industry performs its own risk assessments to demonstrate why they oppose those developed by regulatory agencies. The population exposed to commodities such as household products can be substantially greater at higher exposure levels than the population exposed to most pollutants from industrial sources. The objective of these risk assessments is to uncover any possible source of toxicity that would taint the reputation of the product; hence, this kind of risk characterization has a different flavor from those involving environmental pollutants whose control is likely to impact industrial practices.

II. HISTORICAL ASPECTS OF CARCINOGEN RISK CHARACTERIZATION

Historically, the EPA began risk assessment in the cancer area requiring the initial assessment to indicate whether there was enough basis to launch a full-scale investigation of an agent as a carcinogen. Not much evidence was needed. This was the hair trigger approach (EPA 1976; Albert et al. 1977). At that time, the risk characterization was nothing more than a statement (e.g., there was "significant" evidence for carcinogenicity).

During the 1980s, there was a strong antiregulatory backlash and it seemed appropriate for a number of reasons to qualify the strength of evidence for carcinogenicity (Albert 1985). This involved a stratification of the evidence for carcinogenicity in terms of a letter grade (A for definite, B1 for highly probable, B2 for probable, and C for possible). The risk characterization section consisted of a joint presentation of the grade of carcinogenicity together with a potency factor, the unit risk, for use in estimating population risk by multiplication with the level of exposure. At that time, the EPA's risk assessments were being done by the Carcinogen Assessment Group (CAG). There was no exposure assessment group and, in fact, exposure assessment in those days was primitive. The situation has since improved so that current risk assessments include exposure assessment.

In its original guidelines, the EPA advocated the use of several mathematical extrapolation models, although it was realized that the cornerstone of quantitative risk assessment would become the linear nonthreshold dose–response model. This occurred because there was a strong impetus toward regulating carcinogens as a means of reducing the public health burden of cancer, and the linear model of all the commonly used models provided the highest levels of risk and, thus, the strongest basis for regulation. The linear nonthreshold model means that the risk is proportional to the dose and, most importantly, any dose, however small, can have a calculable excess cancer risk; the risk is zero only for a zero dose. This model had precedent in its use by a federal agency, namely the Atomic Energy Commission, for the estimation of bone and thyroid cancers from radioactive fallout from nuclear testing. The initial approach used by the EPA began by taking the lowest statistically significant dose–response point and drawing a straight line from the 95% upper confidence level of that data point down to zero at the origin of the graph. The slope from the 95% upper confidence limit was called the unit risk (q_i*) and was a measure of the carcinogenic potency of the agent. Later, in response to complaints about throwing away all the data except the lowest response point, the approach was shifted to the multistage model. This model has justification in the multistage concept of cancer as a progression through a series of stages of increasing malignancy. The model assumes that the carcinogen in question has the same action as whatever is causing background cancer (i.e., cancer that occurs in the absence of any known carcinogen exposure). This assumption is the basis for the low-dose linearity of the dose–response curve.

There was always ambivalence about the use of the linear nonthreshold model for nongenotoxic carcinogens. This occurred because the experimental data on tumor promoters, a category of such agents, indicated a threshold-like dose–response pattern and a reversibility of the oncogenic action. This is inconsistent with low level linearity because it would be expected that, at low doses, reversibility (e.g., repair) would dominate and there would be no tumorigenic effect. In formulating its risk assessment guidelines, the EPA was aware of the uncertainty associated with low-level linear risk assessments and took the position that these estimates should be regarded as plausible upper limits of risk (i.e., those which were not likely to be higher, but could be substantially lower). While this action moved the science of risk assessment away from the dilemma of unknowable risks, it put on the risk manager the burden of coping with upper-limit risk estimates. This was difficult to do and, hence, tended to be ignored.

In the 1986 revision of the guidelines (about ten years after the initial "interim" guidelines) (EPA 1986; Albert 1985), the risk characterization section merely called for the presentation of the letter grade of hazard and the slope of the low-dose linear portion of the multistage model—the unit risk. No particular injunctions were given about presentation of uncertainties in the risk assessments, as is the current fashion. Uncertainty weakens the force to regulation and at that time some of the original fervor for control of environmental carcinogens still existed. There were intense arguments about interpretations of results. However, this did not reflect uncertainty; these arguments represented irreconcilable convictions. Nevertheless, the issues did get into the assessments.

III. CURRENT ASPECTS OF CARCINOGEN RISK CHARACTERIZATION

Risk characterization is the component of the risk assessment that produces both the population and the individual risk estimates. It is obtained by multiplying the dose by the probability of response as derived from a dose–response model. The dose can be the average for the population as a whole. This is the simplest to derive, particularly with the linear nonthreshold dose–response model. Nonlinear dose–response models make the calculation more complex because the various dose levels and the number of people involved at each dose have their own probability of response, and the average response is the summation of the risk for the individual dose levels. The maximum level of risk used to be determined by the worst case scenario (e.g., the cancer lifetime risk from arsenic exposure for a person spending his entire life at the boundary fence of the emitting facility). A more sophisticated approach involves the combined probabilities of the important factors that play a role in exposure, each of which has its own probability distribution. The combination of these factors by Monte Carlo methods yields a distribution of exposures, which is advantageous for examining the risks to the most heavily exposed segment of the population, however this is defined (e.g., 90% or 99%)(EPA 1996b). The method is sensitive to the goodness of the distributions of the individual components of exposure and inadequate knowledge of these components can lead to erroneous results. It is not uncommon to have a series of risk estimates presented based on a variety of models. The difficulty is that the various models conform to the data in the observed range but the departure at low doses can involve order-of-magnitude differences.

The practical importance of having a summary section that offers conclusions about the entire health assessment is that there needs to be a bottom line to the assessment. The risk assessment provides the impetus to regulation. The costs of implementation constitute an impediment to regulation. The severity of hazard associated with an agent (i.e., the more grave the effects or potency), and the higher the quantitative risk associated with exposure, the greater the impetus to regulation. The impetus loses force as uncertainty grows in both the hazard and dose–response– exposure assessments. The evaluation must be presented in words to be understand- able. Examples of summary statements with progressively diminishing force for regulation are the following:

1. This is an unequivocal and potent carcinogen with widespread exposure that is now causing large increases in cancer deaths.
2. This is a respiratory irritant that reduces resistance to respiratory infection in children, and good and extensive exposure and epidemiological studies indicate that current indoor exposure levels are producing significant health damage.
3. This agent appears to be a potent carcinogen, but the data are limited by few and inadequate biomedical and exposure studies.
4. This is an agent with equivocal carcinogenicity, but widespread and well-docu- mented exposure that might produce a measurable number of cancer deaths at current exposure levels.
5. This is a mixed aerosol correlated with episodic mortality surges; the association is controversial and the biological rationale for the association is obscure, but the data involve large effects on terminally ill populations.
6. This is a physical agent that is associated with cancer in children in a large number of epidemiological studies, of which about half are positive; the measured expo- sures are not well correlated with cancer and there is at present no biological plausibility to the association.

The summary of the risk characterization section is for use by risk managers who have the decision-making responsibility for regulation or control. Risk managers are not generally trained in health matters. The summary section is what they will focus on and it needs to be stated clearly and nontechnically. From the standpoint of the risk manager, the less uncertainty the better (e.g., is it a carcinogen and, if so, how many people will it harm?). The risk manager has much to deal with in working out whether and how to regulate or control, and the more uncertainty from the biomedical standpoint the more vulnerable the regulator is to the inevitable attack, legal or otherwise, on its proposed regulations and controls. However, since the biomedical basis for both the qualitative and quantitative risk assessment is rarely straightforward, it is necessary to present the uncertainties in the assessment. There is nothing wrong with the concept of risk assessment as a process. It is a valuable method of presenting and analyzing, in a systematic way, the available toxicological and exposure information. Difficulties arise from data gaps and default assumptions.

There are two categories of uncertainty that need to be dealt with in a risk assessment summary:

1. Generic uncertainties that arise from lack of knowledge of the basic biological processes including those that underlie dose–response relationships particularly at low levels of exposure.
2. Uncertainties that are particular to the risk assessment at hand in terms of the quality and scope of the data, and issues that need to be settled as a matter of policy (e.g., should benign tumors be included with cancers in estimating the risk?).

The demands for documentation of uncertainty in risk assessments have increased markedly over the last decade. Why this occurred is not clear. Possibly the scientific controversies over specific risk assessments have been so great that both the scientific and general public have become uneasy about risk assessments and, therefore, regulatory agencies have become more assiduous in documenting uncertainties to promote scientific integrity. Perhaps it is to defuse those who are regulated who would raise all of these uncertainties themselves in objecting to the regulation. Or, it may be the revenge of the risk assessors on the risk managers who tell them what to assess, give them impossible deadlines for doing so, and then have all the fun of calling the regulatory shots, which they, in fact, have been known to avoid until sued.

In the regulatory arena, this territoriality is the so-called risk assessment-risk management paradigm promulgated by the NRC, which places health professionals who do the risk assessments in the position of serving the risk managers. This paradigm is actually a formalization of the existing organizational framework in the EPA, and this is the consequence of the way many offices of the agency were formed as a result of separate pieces of Congressional legislation over many years. The EPA's Office of Research and Development (ORD) is the scientific arm of the EPA and the leader of the risk assessment activities in the Agency. However, the regulatory philosophies in the various laws dealing with risk assessment are different for the different offices. For example, pesticide legislation weighs risks and benefits; air pollution legislation protects everybody with a margin of safety, which in some areas involves technological feasibility with adjustment for residual risk; and water pollution legislation requires the best available technology. The EPA's Office of Radiation and Indoor Air, which handles regulatory activities on radiation, is unique in its interaction with powerful and independent groups like the National Council for Radiation Protection, the International Commission on Radiation Protection, the International Atomic Energy Agency, and the Nuclear Regulatory Commission.

There is agreement in principle that risk assessment should be performed independently of risk management in order to avoid political influence. However, several regulatory offices in the EPA developed their own risk assessment groups independent of the central assessment group in ORD; this was done as a matter of agency policy to decentralize risk assessment in the 1980s. Why this was done is not clear. It may have been a matter of bureaucratic territoriality, a desire to have risk assessment under the control of risk managers, or a need to have experts on immediate call to deal with risk assessment issues. In any case, it is appropriate, in evaluating risk assessments, to note who performed them. The National Institute for Occupational Safety and Health conducts risk assessments independent of their regulatory counterpart, the Occupational Safety and Health Administration (OSHA), but these assessments are unsolicited and advisory, and are frequently ignored; OSHA does its own assessments.

The strengths and weaknesses of the exposure assessment need to be discussed, and of particular concern is the relevance of the exposure route to the risk estimate. The exposure assessment is frequently the weakest part of the risk assessment because of poor analytic methodology, inadequate sampling strategy, or lack of thoroughness of the characterization.

The strengths and weaknesses of the data underlying the dose–response relationships need to be discussed, even when the agent has been assigned an IARC-type grade. The difficulty with this system is that each of the gradations—definite, probable, and possible—covers a wide range of strength of evidence. There has been concern about the propriety of regulating "possible" carcinogens such as the chlorinated solvents and pesticides, where such agents produce tumors only in the mouse liver and in only one sex. Very important uncertainties from a regulatory standpoint develop over whether a given agent is at the high level of "possible" or a low level of "probable."

IV. NONCARCINOGEN RISK CHARACTERIZATION

The oldest approach to regulation, which long preceded risk assessment, is the use of safety factors, now called uncertainty factors, that are applied to the lowest observed adverse effect level (LOAEL) or the no-observed adverse effect level (NOAEL) to obtain a standard. The uncertainty factors are always multiples of ten, but the number depends on whether the data are observed in animals or humans, as well as upon the quality of the data. If obtained in animals, the NOAEL is assigned uncertainty factors of 100—ten for extrapolation from animals to humans and another ten for possible differences in sensitivity between animals and humans. If the data are obtained in humans, only a factor of ten is used to account for possible differences in sensitivity. With the LOAEL, a factor of 1000 is used to compensate for the fact that it is based on dosage which produces health damage. An additional factor of ten may be applied for inadequate data. The dose corresponding to that obtained by the use of uncertainty factors is called a reference dose, or RfD. Exposures are related to the RfD in terms of ratios (i.e., if the exposure is half the RfD, the ratio is 0.5).

The standards so derived are considered "safe" with no uncertainties involved and with no quantitative risk estimates assigned to them. The EPA, which pioneered carcinogen risk assessment, is still using uncertainty factors for noncarcinogen assessments.

Before the Supreme Court decision on benzene (International Union Department v. American Petroleum Institute, 448 U.S. 607, 1980), OSHA regulated strictly on considerations of technical and economic feasibility. When OSHA wanted to reduce the benzene standard from 10 ppm to 1 ppm, the Supreme Court rejected the proposal on the grounds that the agency did not show how much benefit would accrue with the reduction. Therefore, OSHA now uses risk assessment to make this estimate of regulatory benefit. This development has had a recent and interesting consequence. Because of the Supreme Court's requirement to demonstrate the benefit of regulation, OSHA is now forced by its lawyers to use dose–response relationships to derive risk estimates for noncancer toxicants, as is done for carcinogens.

V. FUTURE DEVELOPMENTS IN RISK CHARACTERIZATION OF CARCINOGENS AND NONCARCINOGENS

The EPA has been working on the second revision of its carcinogen guidelines since 1988; a draft version was released in 1996 (EPA 1996a). The EPA expects to have these guidelines finalized in 1998. The original guidelines in 1976 took about six months to develop and adopt. The first revision approved in 1986 took over a year to finalize. The increase from six months to ten years in developing successive guidelines illustrates the principle that positions in regulatory agencies tend to become stagnant because of precedent and become extremely difficult to change. Furthermore, the proposed changes are not major. The weight-of-evidence stratification in the hazard analysis section has been softened. Instead of the A (definite), B (probable), and C (possible) categories, the A and B are lumped into "known/likely" and the C category is changed from possible to "cannot be determined." This recognizes the tendency to avoid regulating agents that are called "possible carcinogens" because of weak evidence (e.g., single sex, single species, or single organ with high background) and the difficulty that there is an accumulation of agents at the boundary of B and C (i.e., the classification of an agent at the upper level of C or the lower level of B is almost always a regulatory decision).

In the quantitative aspect of risk assessment, there is a partial return to the original position of beginning the downward extrapolation from the lowest statistically significant data point. Now instead of using the multistage model, the data in the observed range will be modeled to obtain a 10% dose–response point and the upper confidence level at that point, as before, will be the basis for the downward extrapolation. If the extrapolation is done with a linear nonthreshold straight line, there is very little difference in the result compared to that obtained by the multistage model. The change is proffered on the ground that the multistage model is speculative and that "truth in packaging" calls for a simpler approach. Be that as it may, the important and unspoken consequence will be a more smooth transition to the nonlinear low dose extrapolation (i.e., extrapolation that entails much lower risks at low doses). The linear multistage model cannot be used for this purpose. This change will accommodate the growing pressure to use nonlinear extrapolation for nongenotoxic carcinogens.

The unit risk will presumably be retained with the linear slope beginning at the 10% response level, which will be little different from the multistage model. More attention will be paid to the descriptive aspects of risk characterization, particularly to the uncertainties.

There is a scenario, different from the NRC paradigm, that might appeal to some, and which would certainly change the character of risk characterization. In such an arrangement, the risk assessors would come to a judgment as to whether, from a public health standpoint, an agent should be regulated given the current levels of health damage; some indication of a target for regulatory control of exposure would also be provided. The risk manager would then determine where the biggest regulatory benefits will be obtained and whether the costs will be acceptable to the stakeholders—those who are regulated, Congress, and the general public. In other words, the risk manager would determine what is "do-able" and what is affordable. If there are large discrepancies between the target and feasible levels of control, the risk assessors and managers could negotiate a compromise. This arrangement would force the risk assessors to produce a decision document that would reach conclusions based on weighing the strengths and weaknesses of the available evidence. This is different from simply cataloging the strengths and weaknesses of the evidence. In any case, given current practices, the risk characterization should be written as if it were a decision document without the decision.

At a more fundamental level, there is a basic flaw in the current approach to risk assessment. It is impossible to measure the shape of the dose–response curve within the background noise of the metric being used to measure toxicity (e.g., background cancer incidence). If the dose–response curve cannot be determined, it cannot be known. There may be biologically based reasons for assuming a particular shape of a dose–response curve but that does not change its speculative nature. If the dose–response cannot be known at low dose levels, then the risk estimates cannot be anything but speculative. When speculation becomes dogma, we move into the realm of faith—which is more the province of religion than science. The only risks that can be measured are those in populations where statistically significant responses are obtained in groups of humans or animals. The risk to the individual in the population can only be described as an average. Even with uniform exposure, the individual risk can range from zero to some positive value, because of differences in susceptibility, so that the average does not mean much to the individual.

One solution to this problem is to eschew setting standards based on either individual or population risk in favor of setting standards within the range of background uncertainty (noise). Every toxic response that is measurable has a background present in the absence of exposure to the toxicant in question. If the standard is within the background noise level, it is smaller than the aggregate of the other causes of the same effect and is statistically nonsignificant. Statistical nonsignificance means that the risk is imperceptible and, therefore, societally insignificant in relation to other, more visible, problems. The possibility that statistically nonsignificant population risks may entail significant risks to individuals may be real, but it is unquantifiable in the absence of information about specially susceptible subpopulations that should, if known, be considered separately. The implementation of such an approach would strike a better balance between individual and population risks; it would focus on societally important problems, and provide a uniform method of setting standards for carcinogens and noncarcinogens and both chemical and physical toxicants.

BIBLIOGRAPHY

Albert R.E. 1985. U.S. Environmental Protection Agency Revised Interim, Guideline for the Health Assessment of Suspect Carcinogens, In: *Banbury Report 19: Risk Quantitation and Regulatory Policy*, Cold Spring Harbor Laboratory, 307.

Albert, R.E. 1994. Carcinogen risk assessment in the U.S. Environmental Protection Agency, *Critical Reviews in Toxicology*, 24:75–85.

Albert, R.E., Train, R.E., et al. 1977. Rationale developed by the Environmental Protection Agency for the assessment of carcinogenic risks, *Journal of the National Cancer Institute* 58:1537–1541.

Consumer Product Safety Commission (CPSC). 1984. Carcinogenic Risk Assessment for Formaldehyde: Risk from Exposure to Low Levels Such as Found in Indoor Air, Consumer Product Safety Commission: Washington, DC.

Dyer, R.S., DeRosa, C.T. 1995. Session summary: Chemical mixtures—defining the problem, *Toxicology*, 105:109–110.

Environmental Protection Agency (EPA). 1976. Interim procedures and guidelines for health risk and economic impact assessments of suspected carcinogens, *41 FR 21402*.

Environmental Protection Agency (EPA). 1986. Guidelines for carcinogen risk assessment, *51 FR 33992*.

Environmental Protection Agency (EPA). 1996a. Proposed Guidelines for Carcinogen Risk Assessment, Report No. EPA/600/P-92/003C, Office of Research and Development, U.S. Environmental Protection Agency.

Environmental Protection Agency (EPA). 1996b. Summary Report for the Workshop on Monte Carlo Analysis, Report No. EPA/630/R-96/010, U.S. Environmental Protection Agency: Washington, DC.

Feron, V.J., Woutersen, R.A., et al. 1992. Indoor air, a variable complex mixture: Strategy for selection of (combinations of) chemicals with high health hazard potential, *Environmental Technology* 13:341–350.

International Agency for Research on Cancer (IARC). 1987. *Monograph on the Evaluation of Carcinogenic Risks to Humans, Supplement 7, Overall Evaluations of Carcinogenicity: An Updating of IARC Monographs Volume 1 to 42*. World Health Organization, International Agency for Research on Cancer: Lyon, France.

Janko, M., Gould, D.C., et al. 1995. Dust mite allergens in the office environment, *American Industrial Hygiene Association Journal* 56:1133–1140.

Kaplan, M.P., Brandt-Rauf, P., et al. 1993. Residential releases of number 2 fuel oil: A contributor to indoor air pollution, *American Journal of Public Health* 83(1):84–88.

Keeney, R.L., von Winterfeldt, D. 1986. Improving risk communication, *Risk Analysis* 6(4):417–424.

National Research Council. 1983. *Risk Assessment in the Federal Government: Managing the Process*, National Academy Press: Washington D.C.

National Research Council. 1994. *Science and Judgment in Risk Assessment*, National Academy Press: Washington, D.C.

Nexo, B.A. 1995. Risk assessment methodologies for carcinogenic compounds in indoor air, *Scandinavian Journal of Work Environmental Health* 21:376–381.

Plough, A., Krimsky, S. 1987. The emergence of risk communication studies: Social and political context, *Science, Technology, and Human Values* 12(3/4):4–10.

Repace, J.L., Lowrey, A.H. 1993. An enforceable indoor air quality standard for environmental tobacco smoke in the workplace, *Risk Analysis* 13(4):463–475.

Richmond, H.M. 1991. Overview of a Decision Analytic Approach to Noncancer Health Risk Assessment, for presentation at 84th annual meeting and exhibition of the Air and Waste Managment Association, June 16–21, 1991.

Rothman, A.L., Weintraub, M.I. 1995. The sick building syndrome and mass hysteria, *Neurologic Clinics* 13(2):405–412.

Russell, M., Gruber, M. 1987. Risk assessment in environmental policy-making, *Science* 236:286–290.

Slovic, P. 1986. Informing and educating the public about risk, *Risk Analysis* 6(4):403–415.

Slovic, P. 1987. Perception of risk, *Science,* 236:280–290.

Stolwijk, J.A.J. 1990. Assessment of population exposure and carcinogenic risk posed by volatile organic compounds in indoor air, *Risk Analysis* 10(1):49.

Stolwijk, J.A.J. 1992. Risk assessment of acute health and comfort effects of indoor air pollution, *Annals of the New York Academy of Sciences* 641:56–62.

Wilson, R., Crouch, E.A.C. 1987. Risk assessment and comparisons: An introduction, *Science* 236:267–270.

CHAPTER 7

Characterization of Uncertainty

Steave H. Su, Robert M. Little, and Nicholas J. Gudka

CONTENTS

As far as the laws of mathematics refer to reality, they are not certain; and as far as they are certain, they do not refer to reality.

Albert Einstein (1879–1955)

I. INTRODUCTION

Uncertainty in risk assessment denotes the lack of precise characterization of risk. While the potential for health risks due to exposure to environmental pollutants is known, the level of risk cannot be precisely ascertained — it can only be estimated. For example, the estimates of excess cancer risk from exposure to volatile organic chemicals (VOCs) emitted from building materials can be highly uncertain. This uncertainty has many origins: the emission rates of VOCs are difficult to characterize; the individual's time in the building is variable; and the toxic potentials of the chemicals are uncertain. For this example, the estimated risk can differ by orders of magnitude under different assumptions of exposure and physicochemical parameters. Such degree of uncertainty in any risk assessment is not surprising. Under the current risk assessment methodology the estimated risks are expected to contain uncertainty spanning an order of magnitude or more as a result of the uncertainties associated with the underlying elements.

Since the inception of the current risk assessment paradigm, scientists and policy makers have stressed the need to address the uncertainties inherent in risk assessment (NRC 1983; EPA 1992a; NRC 1994; Commission on Risk Assessment and Risk Management 1997). Despite recognizing that uncertainty should be addressed, there has been limited interest in the regulatory agencies and, thus, minimal guidance to risk assessors. For example, in the 1989 risk assessment guidance for Superfund sites (EPA 1989), the EPA recommended qualitative and semiquantitative characterization of uncertainty since "highly statistical" uncertainty analysis was deemed "not practical or necessary." As a result, past approaches to address uncertainty in risk assessment usually involved using margins of safety or assuming conservative scenarios. These approaches were deemed necessary in order to protect public health; however, these approaches sometimes lacked an adequate scientific basis and, more importantly, provided inadequate characterization of uncertainty. Without a proper characterization of uncertainty, risk assessments often result in excessively conservative estimated risk that is unrealistic (Bogen 1994). Better characterization of uncertainty is necessary because a poor characterization can lead to adverse impact on public health or impractical environmental policy and regulation due to "false sense of certainty" (NRC 1994). Improved characterization of uncertainty will also help focus scientific resources on areas that will reduce major uncertainties in risk assessment. In recent years, public health scientists and policy makers have recognized that better characterization of uncertainty is a more appropriate approach to address uncertainty. Specifically, there is a growing focus on quantifying uncertainties and assessing their impacts on the risk assessment process (EPA 1992a; NRC 1994; Morgan and Henrion 1990).

This chapter provides a discussion of how uncertainty arises in risk assessment. The discussion will be followed by scientific descriptions of the types of uncertainty. There will be an overview of how uncertainty has been treated in the regulatory framework. The chapter will then provide an account of methods recommended to improve the characterization of uncertainty in risk assessment. Finally, a brief overview will be provided of issues regarding the communication of uncertainty.

II. UNCERTAINTIES IN RISK ASSESSMENT

The National Research Council (NRC) described uncertainty in risk assessment as a problem that is large, complex, and nearly intractable (NRC 1994). Uncertainty in risk assessment is too pervasive to describe every instance in which it can arise. However, a review of some of the issues that may arise in each of the four steps of risk assessment, defined in Chapter 2, will help illustrate how uncertainty may spawn in risk assessment. Table 7.1 provides a useful summary of major sources of uncertainty in the current framework of risk assessment.

A. Uncertainty in the Four Steps of Risk Assessment

1. Hazard Identification

Hazard identification examines whether human exposure to an environmental agent has the potential to cause a toxic response or increase the incidence of cancer (EPA 1986; EPA 1996a). For most environmental agents, the human health effects of low-dose, long-term exposure to these agents are uncertain because available data usually do not include results from well-conducted epidemiological studies. In most instances, the potential for an environmental pollutant to be a human carcinogen is determined via results of animal studies. It is questionable (i.e., uncertain) whether carcinogenicity found in animals allows us to assume human carcinogenicity given the physiological differences between species (Calabrese 1987). The issue of the applicability of animal data also raises questions on the appropriate types of animal model (e.g., species, exposure routes, and exposure duration). In some instances, results from animal studies conflict with one test species indicating carcinogenicity while another test species does not. Where human epidemiology data are available, there can still be critical uncertainties. Additionally, difficulty in determining the positive or negative association of exposure and disease incidence can create uncertainty in identifying hazards.

Table 7.1   Major Sources of Uncertainty in Risk Assessment

Hazard Identification	Dose–Response Assessment	Exposure Assessment	Risk Characterization
Different study types: Prospective, case-control, bioassay, *in vivo* screen, *in vitro* screen Test species, strain, sex, system Exposure route, duration	Model selection for low-dose risk extrapolation Low-dose functional behavior of dose–response relationship (threshold, sublinear, supralinear, flexible) Role of time (dose frequency, rate, duration, age at exposure, fraction of lifetime exposed) Pharmacokinetic model of effective dose as a function of applied dose Impact of competing risks	Contamination scenario characterization (production, distribution, domestic and industrial storage and use, disposal, environmental transport, transformation and decay, geographic bounds, temporal bounds Environmental fate model selection (structural error) Parameter estimation error Field measurement error	Component uncertainties Hazard identification Dose–response assessment Exposure assessment
Definition of incidence of an outcome in a given study (positive-negative association of incidence with exposure)	Definition of "positive responses" in a given study Independent vs. joint events Continuous vs. dichotomous input response data	Exposure scenario characterization Exposure route identification (dermal, respiratory, dietary) Exposure dynamics model (absorption, intake processes)	
Different study results Different study qualities Conduct Definition of control populations Physical-chemical similarity of chemical studied to that of concern	Parameter estimation Different dose–response sets Results Qualities Types	Integrated exposure profile Target population identification Potentially exposed populations Population stability over time	
Unidentified hazards	Extrapolation of tested doses to human doses		
Extrapolation of available evidence to target human population			

Adapted from Bogen (1990).

2. Dose–Response Assessment

Dose–response assessment examines the relationship of dose to the degree of response observed in an animal experiment or human epidemiological study. Like hazard identification, incomplete toxicity information drives uncertainty in dose–response assessment; however, dose–response assessment is quantitative and any uncertainty is unavoidably incorporated into its calculations. Consequently, the amount of uncertainty in a dose–response relationship is highly dependent on each chemical's toxicity database. For example, a few chemicals (e.g., arsenic) have sufficient epidemiological data of occupational cohorts for the EPA's derivation of a dose–response relationship (carcinogenic slope factor) but, more frequently, animal data are used to derive a dose–response relationship. In the dose–response assessment of a carcinogen, three extrapolations are frequently needed: (1) from high to low dose, (2) from animal to human responses, and (3) from one route of exposure to another (EPA 1986). When exposure–response data are obtained from animal studies, there are questions on the appropriate dosimetric scaling to reflect a human-equivalent dose (EPA 1992b). In addition, interhuman variability in pharmacokinetic and pharmacodynamic parameters also presents an uncertainty in dosimetry evaluation; there are also questions about whether the toxicity of chemical mixtures can be characterized based on the toxicity of individual compounds. Finally, one of the greatest sources of uncertainty in risk assessment is the use of mathematical models to extrapolate dose–response data obtained from high-dose experiments to predict response from low doses associated with human exposure (NRC 1983; Beck et al. 1989). Figure 7.1 illustrates that cancer risk predicted from various types of low-dose extrapolation models can differ by orders of magnitude (NRC 1983). The uncertainty in low-dose extrapolation involves issues of whether an exposure threshold exists for carcinogenic effects, and what is the shape of the dose–response curve at low-dose ranges that are not experimentally observable.

3. Exposure Assessment

In the exposure assessment step, uncertainties arise from the inherent difficulty to characterize fully and accurately exposure in the population of concern. The modeling of the fate and transport of environmental pollutants often presents a challenge in exposure characterization (NRC 1991). In developing mathematical models that describe transport of pollutants from their source to human receptors, uncertainties result from unrealistic characterization of source release, physicochemical interaction with the environmental media, and other relevant parameters. Uncertainties also arise during characterization of human activities and physiological parameters related to exposure (Whitmyre et al. 1992). In developing exposure scenarios, uncertainties include whether individuals may enter the microenvironments where pollutants exist, the frequency of such events, and the duration. There also is uncertainty in characterizing the physiological process of intake of the pollutants, which include respiration rate and dermal and gastrointestinal absorption efficiency.

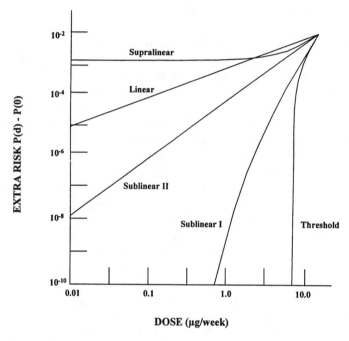

Figure 7.1 Uncertainty of estimating cancer risk with low-dose extrapolation models.

4. Risk Characterization

Quantification of a risk estimate is achieved by combining the results of the exposure and dose–response assessments to produce an estimate of risk to the individual (i.e., hazard quotient for noncancer effects and excess lifetime risk for cancer). Consequently, the uncertainties in quantification of risk estimates are a result of the earlier steps of the risk assessment (i.e., interpretation of hazard identification, assumptions in dose–response relationship, or incomplete exposure characterization). The uncertainties associated with each of the three steps may combine and propagate the overall uncertainty. Another source of uncertainty in risk characterization is determining which substances and pathways involve similar modes of actions and should have their risks summed (EPA 1989). The final risk characterization may be highly uncertain and the estimated risk may span several orders of magnitude.

III. DEFINING UNCERTAINTY

Uncertainty is a general term indicating the lack of precision in an estimated quantity (i.e., cancer risk for an exposed population). To address uncertainty in risk assessment it is useful to define this rather abstract terminology. A more refined description of uncertainty separates it into two categories: (1) variability, referred to as Type A Uncertainty, and (2) uncertainty, or Type B Uncertainty (Hoffman and Hammonds 1994; IAEA 1989). In the Guidelines for Exposure Assessment (EPA 1992c) and Guidance for Risk Characterization for Risk Managers and Risk Assessors (EPA 1992a), the EPA advised that the two types of uncertainty be clearly distinguished. The National Research Council (1994) and Commission on Risk Assessment and Risk Management (1997) also urge the distinction between uncertainty and variability. Separate characterization of uncertainty and variability will help distinguish between uncertainty that can be reduced and variability that must be accepted (EPA 1992c).

A. Variability

Variability denotes the heterogeneity in nature and is associated with an inability to generalize a parameter using a single number. Any attempts to describe a parameter of this type (e.g., body weight) with a single number will fail to describe its distribution (e.g., the range of body weights in the population). This can result in over- or underestimation of risk for the entire population, as well as failing to provide a measure of the range of risks to individuals. When variability is well characterized from survey analysis, additional scientific study will better characterize this variability, but will not eliminate it.

The EPA has defined three types of variability in the Draft Exposure Factors Handbook (EPA 1996b): (1) spatial variability, (2) temporal variability, and (3) interindividual variability. Spatial variability represents variability across locations at a local (micro) or regional (macro) scale. An example of spatial variability would be the differences in air concentration of respirable suspended particles in different areas within a home. Temporal variability represents variability over time, whether long-term or short-term. An example of temporal variability would be seasonal differences in air exchange rates for a home. Interindividual variability represents the heterogeneity in a population. Individuals in a population differ in their physiological parameters as well as in their behavior (e.g., body weights, time spent inside their home, etc.).

B. Uncertainty

Uncertainty denotes the lack of precision due to imperfect science. It differs from variability in that uncertainty can be reduced with improved science (e.g., better devices or methods). An example of this type of uncertainty is the determination of the speed of light. The determination of the speed of light over the history of science evolved from early, crude estimates that were highly uncertain, to recent, more precise measurements (Morgan and Henrion 1990).

It is helpful to define uncertainty by classifying it into three broad categories: (1) scenario uncertainty, (2) parameter uncertainty, and (3) model uncertainty (EPA 1992c; EPA 1996b). Scenario uncertainty represents uncertainty due to missing or incomplete information needed to totally describe a scenario. Parameter uncertainty represents uncertainty in parameters that are measured or estimated. Model uncertainty

Table 7.2   Scenario, Parameter, and Model Uncertainty (Type B Uncertainty)

Type of Uncertainty	Sources	Examples
Scenario uncertainty	Descriptive errors	Incorrect or insufficient information
	Aggregation errors	Spatial or temporal approximations
	Judgment errors	Selection of an incorrect model
	Incomplete analysis	Overlooking an important pathway
Parameter	Measurement errors	Imprecise or biased measurements
uncertainty	Sampling errors	Small or unrepresentative samples
	Variability	In time, space, or activities
	Surrogate data	Structurally related chemicals
Model uncertainty	Relationship errors	Incorrect inference of the basis for correlation
	Modeling errors	Excluding relevant variables

Adapted from EPA 1996b

represents the inability of models to represent thoroughly the real world. Table 7.2 summarizes the sources and some examples of these three types of uncertainty.

C. Other Frameworks

Uncertainty may be defined in other frameworks. Some scientists prefer to partition uncertainty into three categories: (1) bias, (2) randomness, and (3) true variability (NRC 1994; Hattis and Anderson 1993). In this framework, *bias* is the uncertainty resulted from study design and performance, *randomness* is the uncertainty due to sample size and measurement imprecision, and *true variability* is the uncertainty associated with heterogeneity in nature. Morgan and Henrion (1990) also provided a useful framework to characterize uncertainty.

IV. THE EPA APPROACH TO ADDRESSING UNCERTAINTY

As discussed in Section II, uncertainty is present in each step of the risk assessment process. Although the need to characterize uncertainty has been evident since the risk assessment process was formalized, the guidance provided by the EPA, until recently, was limited. The general approach used by the EPA in the past involved either qualitative discussion of uncertainty or conservative quantitative estimates. The following discussion will cover the primary means that EPA regulations and guidelines use to handle uncertainty in the four steps of a risk assessment.

A. Hazard Identification

Because there is a lack of human data establishing carcinogenicity for most chemicals, the EPA relies on the results of animal models, *in vitro* toxicity tests, and, to a limited extent, structure-activity relationships (EPA 1986; EPA 1992d; EPA 1996a). Use of these alternative data sources due to the absence of human data represents a major uncertainty. To address this uncertainty, the EPA developed the categorization scheme described in Chapter 2 to classify the carcinogenicity of chemicals based on a weight-of-evidence approach (Group A, B1, B2, C, D, E). For example, a chemical shown to be carcinogenic to rats or mice under high dose, lifetime (2-year) exposure, would be classified as a probable human carcinogen (B2), and treated as a carcinogen in risk assessment even without supporting human epidemiological data.

For noncancer effects (e.g., neurotoxicity and hepatotoxicity), the EPA's determination of potential human toxicity also relies on the weight-of-evidence approach with an emphasis on animal models (EPA 1992d). Just as in the carcinogen assessment, if a chemical is found to be toxic to animal species, similar effects in humans are assumed. In addition, humans are presumed to be more sensitive to toxicity than animals so that the uncertainty of animal-to-human extrapolation of toxic effects is treated via an "uncertainty factor" as described in the next section.

B. Dose–Response Assessment

The derivation of a dose–response relationship contains many uncertainties from extrapolation between species, routes, and high-dose to low-dose exposure. The EPA handles the uncertainty of extrapolating dosimetry from animal exposure to human exposure by deriving the human equivalent dose using a scaling scheme based on body weight of the animal species (EPA 1992b). A contentious issue in dose–response assessment of a carcinogen is high- to low-dose extrapolation. While the possibility of a toxic threshold for carcinogens is still being debated among scientists, the EPA has taken a conservative approach to address this uncertainty by assuming there is no threshold to any carcinogen (i.e., a carcinogen can cause cancer at any dose), meaning the dose–response curve originates from the zero dose (EPA 1986; EPA 1996a; Melnick et al. 1996). Furthermore, the shape of the dose–response curve at low doses associated with environmental exposure is unknown. Many dose–response models are available and they can predict vastly different responses (i.e., cancer risk). Figure 7.1 shows how four dose–response models applied to the same set of data predict dramatically different risks. The EPA default approach is to assume the curve at low dose is linear with the potency of the carcinogen determined by the slope. Furthermore, a conservative approach is utilized to derive the carcinogenic slope factor given the limited amount of data points characterizing the dose–response relationships. From the dose–response model, the statistical upper-95th percent confidence limit of the estimated slope factor is used as the cancer potency factor in risk assessment.

The EPA describes the uncertainty in the dose–response assessment of chemicals by stating a level of "confidence." This discussion of confidence describes the ability of the risk values derived from dose–response assessment on the agent to estimate the risks of that agent to humans (EPA 1992d). This judgment is based on the consideration of factors that increase or decrease confidence in the numerical risk estimate. The confidence statements, however, are of a qualitative nature and do not represent any quantitative characterization of the uncertainty surrounding the derivation of the dose–response relationship.

The EPA's assessment of noncarcinogenic effects assumes that there is a threshold to toxic effects. This means there is a range of exposure from zero to the threshold that can be tolerated by the organism with essentially no chance of expression of adverse effects (EPA 1989). The EPA approach to noncarcinogens involves the development of an oral reference dose or inhalation reference concentration (RfD/RfC) from the no-observed-adverse-effects-level (NOAEL) for the most sensitive, or critical, toxic effect. This is based in part on the assumption that if the critical toxic effect is prevented, then all toxic effects are prevented (EPA 1989; EPA 1994). This approach also assumes that humans are more sensitive to toxic effects than is the most sensitive animal species tested (EPA 1989). To address the uncertainties involved in deriving the RfD or RfC, the EPA uses uncertainty factors of 10 to account for each of the following: interindividual differences in susceptibility; extrapolation from animals to humans; extrapolation of results from subchronic exposure studies to chronic exposure studies; and lowest-observed-adverse-effect-level (LOAEL) to NOAEL extrapolation. A NOAEL (or LOAEL) is divided by all applicable uncertainty factors and a modifying factor between 1 to 10 (default value = 1) to reflect the professional judgment of the assessor to derive a RfD or RfC (Dourson and Stara 1983; EPA 1989; EPA 1994). Furthermore, for the derivation of RfCs, additional dosimetric scaling of the NOAEL is necessary in order to address the morphological differences in the respiratory systems between experimental animals and humans (EPA 1994).

C. Exposure Assessment

The EPA methodology for conducting exposure assessments has been dictated to a large degree by the substantial level of uncertainty inherent in these assessments. The traditional EPA approach was based on assessing exposure according to two criteria: (1) exposure of the total population, and (2) exposure of a specified, usually highly or maximally exposed, individual (MEI) (NRC 1994). The MEI was supposed to represent a potential upper bound in this old approach; consequently, its calculation was based on numerous conservative assumptions (NRC 1994). One of the more conservative and contentious of these assumptions regarded the target-population identification. Using the EPA's approach, the MEI was assumed to spend 24 hours/day for 365 days/year during a lifetime of 70 years at the location determined by dispersion modeling or field sampling to receive the heaviest annual average concentration with no allowance made for time spent indoors or away from home (EPA 1989).

The EPA recently began considering both a high-end exposure estimate (HEEE) and a theoretical upper-bounding estimate (TUBE). The HEEE and TUBE are designed to work in tandem, with the TUBE providing the upper-bound estimate and the HEEE providing a conservative, but realistic, estimate of actual exposure. The TUBE is used for bounding purposes only and is to be superceded by the HEEE in detailed risk characterizations (NRC 1994). The TUBE was designed to be an easily calculated upper bound by simulating exposure, dose, and risk levels exceeding the levels experienced by all individuals in the actual distribution (NRC 1994). Calculating the TUBE involves using the upper limit for all parameters in the exposure characterization and exposure–dose assessments, as well as the dose–response relationships (NRC 1994). The HEEE was designed to serve as a plausible exposure estimate to individuals at the upper end of the exposure distribution (i.e., above the 90th percentile of the population, but not higher than the individual with the highest exposure distributions). The HEEE was intended to replace the combination of average and upper-bound case (the previous approach) as a decision making tool because it is more realistic than an upper-bound exposure estimate, while more protective in light of uncertainty than an average exposure estimate (EPA 1989).

While the MEI and TUBE use the stringently conservative assumptions to incorporate uncertainty into their upper-bound determination, the HEEE uses different assumptions about contamination-scenario characterization, exposure-scenario characterization, target-population identification, and integrated exposure profile to develop a conservative, but plausible, exposure estimate. The reasonable maximum exposure (RME) is a HEEE which is the basis for all actions at Superfund sites (EPA 1989). To address uncertainties in the Superfund risk assessment, the guidance has deemed the statistical 95th percent upper confidence limit as appropriate for several key parameters for the RME's exposure characterization (EPA 1989). For example, the guidance requires the use of the 95th percentile of exposure concentration, contact rate (i.e., amount of contaminated medium contacted per unit time or event), and exposure frequency and duration for calculation of the RME when such data are available (EPA 1989).

D. Risk Characterization

Currently, the final risk estimates from risk assessments are presented as deterministic estimates. These risk estimates are not accompanied by any quantitative description of the uncertainty surrounding the value. The prevailing EPA approach to characterize uncertainty in risk characterization is to describe sources of uncertainty individually, in qualitative or quantitative terms (EPA 1989). By describing only the sources, however, there is no quantitative characterization of the imprecision of the risk estimate. In addition, the discussion of sources of uncertainty individually does not allow one to assess the effect of uncertainties propagated through the risk assessment process. The EPA's Guidance on Risk Characterization for Risk Managers and Risk Assessors (EPA 1992a) suggests combining ranges of exposure estimates to provide "multiple risk descriptors." However, there is limited implementation of this recommendation throughout the EPA programs and offices.

V. RECOMMENDED METHODS TO CHARACTERIZE UNCERTAINTY

This section will discuss the methods recommended to characterize the uncertainties in risk assessment. Many of these methods are dictated by the EPA guidelines. In areas where specific guidelines are lacking, the methods presented will reflect state-of-the-art approaches to characterize uncertainty.

Before discussing uncertainty characterization for each of the four components of risk assessment, it is useful to envision an ideal uncertainty characterization. Uncertainty, when characterized to show both uncertainty and variability, allows decision makers to identify the magnitude of estimated risk attributed to different segments of the exposed population (e.g., high-end or low-end) and the uncertainties associated with the estimates. The NRC (1994) presented the ideal uncertainty characterization of risk assessment that separated uncertainty from variability. Figure 7.2 shows a cumulative distribution plot of risk vs. population percentile with confidence bounds (NRC 1994). The solid line indicates the most likely distribution of risks across the exposed population and is indicative of the variability in the estimated risk. The dashed lines that envelope the most likely estimates show the upper- and lower-bounds of the estimated risk. The upper- and lower-bounds are indicative of the uncertainty surrounding the estimated risk across the exposed population. A similar recommendation for characterizing both uncertainty and variability for the purpose of exposure assessment was presented by the EPA (EPA 1992c).

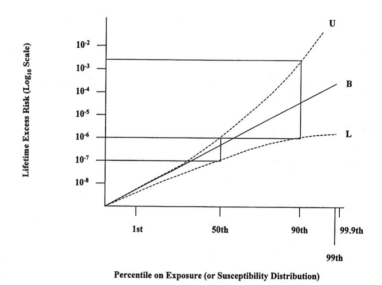

Figure 7.2 Separate characterization of uncertainty and variability.

While the NRC presented the ideal characterization of uncertainty, practical considerations make achieving such a goal difficult. In particular, the type of uncertainty characterization described by the NRC would require quantitative uncertainty analysis throughout each of the four components in risk assessment. An obstacle to fully characterizing uncertainty in the current risk assessment process is the historical reliance on deterministic (i.e., single value) estimations of parameters and risk. In addition, current EPA policy precludes the use of quantitative uncertainty analysis in the hazard identification and dose–response steps of risk assessment (EPA 1997b). Presently, the approach to quantitative uncertainty analysis endorsed by the EPA is allowed for use only in exposure assessment and the final risk characterization.

A. Hazard Identification

Currently, the accepted methods to determine potential carcinogenicity in humans of environmental pollutants are provided in the EPA 1986 Guidelines on Carcinogen Assessment (EPA 1986). Under these guidelines, uncertainty in determining human carcinogenicity is addressed in a qualitative and descriptive manner. As noted earlier, chemicals are grouped into categories describing likelihood for carcinogenicity based on available data. The classification of chemicals depends on the weight of evidence from experimental data obtained from different test systems; tumor findings in animals and humans are the dominant component of the decisions. In the EPA's presentation of carcinogen assessment, the uncertainty associated with carcinogenicity determinations is also presented in a qualitative discussion regarding the EPA's "confidence" in the assessment (EPA 1992d).

The EPA's proposed Guidelines for Carcinogen Assessment (EPA 1996a) do not deviate from the qualitative treatise of uncertainties associated with carcinogenicity determination. However, the simplified classification descriptors that will replace the current letter designation with the more descriptive terms "known/likely," "cannot be determined," or "not likely" is similarly based on the weight-of-evidence approach. The proposed guideline, however, encourages the characterization of carcinogenicity by considering a greater scope of experimental as well as the model-derived results. Despite its qualitative nature, the evaluation of a greater range of carcinogenicity-related data improves the characterization of uncertainty.

The expanded scope of carcinogenicity-related data evaluation may create opportunities for quantitative uncertainty assessment in the hazard identification step. Quantitative uncertainty analysis may be applied to mechanistic information from animal or genotoxicity studies used to determine potential human carcinogenicity. An example of using quantitative uncertainty analysis in hazard identification is an approach that was developed to predict animal carcinogenicity from short-term genotoxicity tests. In this type of analysis, the probability of animal carcinogenicity is characterized from results of genotoxicity assays using Bayesian statistics (Chankong et al. 1985). The advantage of this analysis is that it provides a quantitative characterization of the uncertainty of carcinogenicity based upon available data, and it indicates how the uncertainty may be reduced with additional data from specific types of assays.

B. Dose-Response Assessment

The characterization of uncertainty in a dose–response assessment of a carcinogen depends on the methods used in the assessment. EPA (1996a) proposes four methods for dose–response assessment of a carcinogen: (1) biologically based models, (2) curve-fitting and point of departure extrapolation with linear analysis, (3) curve-fitting and point of departure extrapolation with nonlinear analysis, and (4) toxicity equivalence factors (TEFs) (EPA 1996a). These approaches address the uncertainties regarding the dose–response relationship below the observable range and where empirical data are limited. The appropriate method for dose–response assessment is dictated by the amount and quality of the data available. Each approach's applicability, protocol (methodology), and treatment of uncertainty will be discussed individually.

When adequate data are available, biologically based models that relate dose and response data in the range of empirical observations are the preferred tools for dose–response assessment. Recently, the EPA has utilized biologically based models to estimate risk at low-dose exposures for some chemicals (EPA 1997a) and opened the door for the use of more biologically based models to address the uncertainties associated with the selection of low-dose extrapolation models (EPA 1996a). Similarly, the uncertainty of dosimetry scaling from animal studies to human exposure is being addressed using more biologically based approaches, such as physiologically based pharmacokinetic (PBPK) models. It is important to note that uncertainties still exist in these more recent, biologically plausible, dose–response modeling approaches; however, these uncertainties can be characterized in a quantitative manner. For example, the uncertainty of PBPK modeling can be evaluated against differences in model structure and parameter (Hattis et al. 1990; Hattis et al.1993; Woodruff et al. 1992) and, for the linearized multistage model, a probability distribution of the estimated carcinogenicity slope factor can be calculated (Crouch 1996).

When the data necessary for the development of a biologically based model are unavailable, the EPA recommends using curve-fitting and point of departure extrapolation (EPA 1996a). In this approach mathematic modeling (e.g., logistic, polynomial, Weibull) is used to fit the empirical data relating dose and response data in the observable range. The dose associated with an estimated 10% increased tumor incidence then is identified from the lower 95% confidence limit on the fitted curve (LED_{10}). The LED_{10} serves as the point of departure for both linear and nonlinear low-dose extrapolation when the dose–response relationship is characterized by the curve-fitting approach as opposed to using a biologically based model.

Low-dose extrapolation based on the assumption of linearity is appropriate for the following cases: when evidence supports a mode of action that is anticipated to be linear, like gene mutation due to DNA reactivity; if the anticipated human exposure falls on the linear portion of an overall sublinear dose–response curve; or as the ultimate science policy default assumption in cases of inconclusive evidence (EPA 1996a). In these cases, the EPA (1996a) proposes linear extrapolation from the point of departure (e.g., LED_{10}) to the origin (i.e., zero dose, zero response) as shown in Figure 7.3. Using the LED_{10} as point of departure for linear extrapolation determines the quantitative carcinogenic risk expressed as the conservative, upper-bound excess probability of an individual developing cancer over his lifetime. The use of the lower confidence limit on dose appropriately accounts for experimental uncertainty in the dose–response relationship (EPA 1996a). This method of linear extrapolation from LED_{10} produces unit risk values that are comparable to those derived from the traditional approach using linearized multistage models.

When a carcinogen has an apparent threshold and there is other evidence for nonlinearity based on mode of action, an assumption of nonlinearity in the low-dose region may be appropriate. The recommended approach for nonlinearity is the use of a margin of exposure analysis rather than estimating the probability of effects at low doses. Like the RfD/RfC approach, the point of departure (e.g., LED_{10}) is divided by uncertainty factors of no less than tenfold each to account for human variability and for interspecies sensitivity differences. The LED_{10} is also divided by the exposure of interest to provide information on how much reduction in risk may be associated

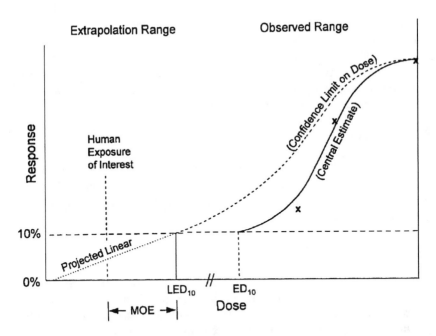

Figure 7.3   Low-dose linear extrapolation of carcinogenicity using LED_{10} as point of departure. *Source:* Adapted from EPA (1996a).

with reduction in exposure from the point of departure. The use of a margin of exposure approach is included as a new default procedure to accommodate cases in which there is sufficient evidence of a nonlinear dose–response, but not enough evidence to construct a mathematical model of the relationship. The use of uncertainty factors (normally 10) to account for the uncertainty associated with human variability and interspecies sensitivity differences creates a conservative estimate sufficiently protective of sensitive populations.

In the event that no acceptable animal or human data are available for a chemical believed to produce effects of toxicological significance, a TEF or relative potency estimate may be used (EPA 1996a). TEFs are used to estimate the toxicity of an unknown compound based on characteristics (e.g., receptor-binding characteristics, results of assays of biological activity related to carcinogenicity, or structure-activity relationships) that are shared with a well-studied member of the same chemical class. TEFs are generally indexed at increments of a factor of 10 with more precise data allowing smaller increments (EPA 1996a). Relative potencies are derived like TEFs but have less supporting data; they are only used when there is no better alternative. The uncertainties associated with TEFs and relative potency estimates should be discussed qualitatively whenever they are used.

C. Exposure Assessment

The current EPA exposure assessment guidelines, the *Guideline on Exposure Assessment* (EPA 1992c) and the *Draft Exposure Factors Handbook* (EPA 1996b), describe some measures to characterize both Type A (variability) and Type B (uncertainty) uncertainties. At the present time, these approaches have not been implemented in the risk assessment processes so there remains a propensity to use conservative assumptions and scenarios in face of uncertainty. For the purpose of providing recommended methods to characterize exposure, it will be useful to discuss the approaches envisioned by these guidelines.

1. Uncertainty

The EPA has classified the uncertainties due to an imperfect state of knowledge (Type B Uncertainty) into three groups: (a) scenario uncertainty, (b) parameter uncertainty, and (c) model uncertainty. The means to address these types of uncertainty are as follows:

a. Scenario Uncertainty

Scenario uncertainty includes descriptive errors, aggregation errors, errors in professional judgment, and incomplete analysis (see Table 7.2). These scenario uncertainties are essentially nonnumeric uncertainties that are not quantifiable. Because of this nonquantifiable nature, scenario uncertainties are best characterized by a qualitative discussion of the rationale behind selecting or formulating specific exposure scenarios. For example, a scenario of workplace exposure would consider only actual working hours rather than the whole week.

b. Parameter Uncertainty

Parameter uncertainty arises from measurement errors, sampling errors, and use of generic or surrogate data. It should be noted that the EPA had included variability (Type A Uncertainty) as one source of parameter uncertainty (EPA 1992c). Since parameter uncertainties involve numeric properties, such uncertainty can be quantifiable. The EPA has suggested several approaches to quantify parameter uncertainties (some of these approaches do not characterize uncertainty well and can lead to conservative interpretations):

1. Order-of-magnitude bounding of the parameter range: This approach provides only a crude estimate of the parameters (e.g., $PM_{2.5}$ emission rate from a woodstove is characterized as between 1 to 10 mg/hr). A significant problem with this approach is that a combination of order-of-magnitude bounding values will result in an estimate that is well below, or well above, the theoretical bounds (e.g., exposure level that is above the TUBE, described above). In addition, such an estimate provides no information on the likelihood that the estimated value will occur.
2. Description of the range of the parameters with lower- and upper-bound values, and best estimates: This is the approach most commonly used in conventional risk assessments. In using such an approach, guidelines may require a specific method to quantify the lower- and upper-bound, the type of data distribution, and the best estimate. The problem associated with this approach is similar to the order-of-magnitude bounding approach. The use of multiple lower-bound or upper-bound

parameter values will result in estimates within the theoretical bounds; however, the estimate may represent highly unlikely exposure scenarios. Furthermore, whether using the lower- or upper-bound, or best estimates of parameters, the final exposure estimate provides no information on the likelihood for the estimated exposure to occur.

3. Sensitivity analysis that changes the value of one variable while holding other variables constant to evaluate the resulting effect on the output: This analysis is useful as a part of screening level analysis since the result will indicate which variable requires further analysis or data gathering.

4. Analytical uncertainty propagation that examines how uncertainty of an individual parameter affects the overall uncertainty of the final estimate: The problem associated with this approach is that determining the necessary mathematical derivative of the exposure equation can be difficult. Also, this approach is most accurate for linear equations and any departure from linearity requires additional evaluation.

5. Classical statistical methods that describe uncertainty by characterizing the distribution of values for each of the exposure parameters: The distribution of values may also be used to calculate confidence intervals of a specific percentile (i.e., uncertainty). The limitation of this approach is that uncertainty is not propagated across all of the model parameters to provide a measure of the total uncertainty of the exposure estimate.

6. Probabilistic uncertainty analysis that uses probability distributions to represent each of the exposure model parameters: The probability distributions indicate all of the possible values that each model variable can hold, and the likelihood of each variable to be any specific value. The EPA has specifically endorsed this type of uncertainty analysis and provided guidance (EPA 1997b). For this reason, probabilistic uncertainty analysis for exposure assessment and its integration with risk characterization will be discussed here at greater length.

The most common form of probabilistic analysis used in risk assessment is Monte Carlo analysis. In Monte Carlo analysis, values are randomly selected from the probability distributions and entered into the exposure equation to obtain an exposure estimate. When this process is repeated many times (i.e., thousands of iterations), the uncertainty of the model parameters are propagated, and the result is a distribution of exposure estimates reflective of the overall uncertainty of the exposure estimate (see Figure 7.4). In addition, recent tools such as @Risk (Palisade Corp 1994) and Crystal Ball (Decisioneering Corp. 1990) also allow sensitivity analysis that characterizes the relative weight of the model variables in contributing to the overall uncertainty. The primary difficulty associated with Monte Carlo analysis is the need to develop appropriate probability distributions for the model parameters and the lack of a single source for all the probability distributions for use in exposure assessment. At the present time, collections of probability distributions may be found in some publications (EPA 1996b; AIHC 1994; Finley et al. 1994). The EPA plans to develop guidance that provides "default" probability distributions (EPA 1997b).

The recent guidance on probabilistic analysis is the first step in the EPA's commitment to quantitative uncertainty analysis, but it does not provide adequate information on practical issues that must be addressed in an actual risk assessment.

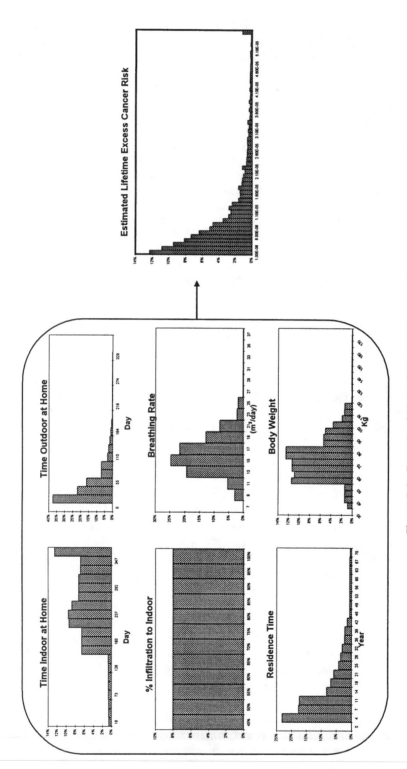

Figure 7.4 Example of Monte Carlo uncertainty analysis.

In addition to the publications that provide useful probability distributions, scientific guidance on "good practices" in Monte Carlo analysis may be found in the paper by Burmaster and Anderson (1994). Some useful examples that illustrate the use of Monte Carlo analysis as well as an alternative probabilistic analysis, such as Bayesian analysis, are provided in McKone and Bogen (1991), Thompson et al. (1992), and Dakins et al. (1994).

c. Model Uncertainty

Model uncertainty arises when more than one conceptual or mathematical model can be used to address the exposure scenario. The EPA advises that a qualitative discussion be made to address the model acceptance by the scientific community and its applicability to the specific problem. The uncertainty of the modeling approach may be addressed by applying the preferred and plausible alternative models and presenting the range of outputs as the uncertainty range. Another type of modeling uncertainty is the uncertain correlations between chemical properties, structure-reactivity correlations, and environmental fate models. An example of correlation uncertainty is individuals changing their breathing rate as a result of high pollutant concentration in air. This type of uncertainty is difficult to characterize since literature data usually focuses on one variable and does not discuss its correlation to other variables.

2. Variability

The EPA (EPA 1996b) has classified the uncertainties due to heterogeneity in nature into spatial variability, temporal variability, and interindividual variability (see Section III); however, there is limited guidance on the actual methods to address these uncertainties other than using conservative estimates (see Section IV). Regarding variability, the EPA and public health scientists have focused on scenarios where a segment of the population is highly exposed due to factors such as pollutant fate and transport, physiological characteristics, and behavioral characteristics. As noted, deterministic approaches, such as the RME and MEI, cannot quantify the proportion of the population segment who are highly exposed. In order to characterize accurately the proportion of high-end exposures, it is necessary to address variability by fully characterizing the distribution of exposure in a population.

The EPA (EPA 1992c) described assessing high-end exposures using a simulated population variability. Such simulation can be achieved using techniques like probabilistic analysis where each exposure parameter is represented by probability distributions. The simulation of population variability using probabilistic analysis may be approached in a manner similar to one recommended for assessing parameter uncertainty (Type B uncertainty). Examples where the population variability is characterized using probabilistic analysis include an assessment of less-than-lifetime exposures to chemical contaminants (Price et al. 1992) and population mobility (Johnson and Capel 1992).

3. Techniques to Separate Characterization of Uncertainty and Variability

The NRC (1994) and the EPA (1992a) stressed the need to characterize separately uncertainty and variability. It is important to note that separate characterization of uncertainty and variability does not imply that the two should be analyzed independently. These analyses should be integrated in order to characterize both uncertainty and variability in a manner envisioned by the NRC (1994). The current EPA guidelines do not suggest a modeling approach that integrates the characterization of uncertainty and variability; however, in recent years approaches have been developed which involve statistical estimation or "nested-loop" Monte Carlo analysis (Bogen and Spear 1987; Bogen 1995; McKone 1994; Frey and Rhodes 1996; Price et al. 1996).

D. Risk Characterization

As discussed in the beginning of this section, risk characterization envisioned by the NRC (1994) provides a quantitative description of estimated risk that indicates both types of uncertainty (Figure 7.2). Producing such a risk characterization requires a full characterization of uncertainty that traditional deterministic risk characterization cannot achieve. It is recommended that quantitative uncertainty analyses be conducted, where possible, in each step of risk assessment. One major obstacle to achieving this goal is the current EPA policy that does not allow quantitative uncertainty analysis in hazard identification and dose–response assessment (EPA 1997b). Currently, the most feasible means is combining the deterministic toxicity potency value with the results of quantitative uncertainty analysis from exposure assessment. Useful examples of treatment of uncertainty of risk characterization may be found in journals such as *Risk Analysis: An International Journal* (Plenum Press, New York) and *Human and Ecological Risk Assessment* (Amherst Scientific Publishers, Amherst).

VI. COMMUNICATION OF UNCERTAINTY IN RISK ASSESSMENT

Risk communication is an important step in the risk assessment process that if handled improperly can render a risk analysis useless, lead to ineffective risk management strategies, and waste scarce resources and attention (Ibrekk and Morgan 1987). Since uncertainty is inherent to risk assessment, reporting uncertainty is an essential part of an accurate risk communication (Johnson and Slovic 1995). The presentation of uncertainty affects how the public perceives risk and, therefore, must be considered carefully.

Traditionally, risk estimates have been presented using point estimates with uncertainty receiving only qualitative treatment, if any. While foreign in the context of science and risk estimates, quantifying uncertainty is actually a familiar part of everyday life. One example is driving to the airport. Estimating the driving time to the airport requires consideration of traffic conditions, such as rush-hour congestion or construction delays, and other factors, such as weather, that can affect driving time. After considering these factors (i.e., uncertainties), one can judge the travel time using a best-case estimate (lower bound), worst-case estimate (upper bound), and most likely travel time. The judgment might be, "The trip should take between an hour to hour and a half, probably an hour and twenty minutes." Other examples of significant everyday uncertainties include weather forecasts, how long to cook food, or the price of a cab ride in Washington, DC.

The public's unfamiliarity with the quantitative presentation of uncertainty in risk assessments makes it difficult to communicate uncertainty effectively. The largest obstacle to reporting uncertainty is simply presenting it in a form which is easily understood. For a technical audience, a histogram, box and whisker, or line plot may suffice; however, care must be taken when presenting uncertainty to the general public. Ibrekk and Morgan (1987) studied a nontechnical audience's responses to nine graphical displays of probabilistic results. They concluded that none of the graphical representations would be clear to everyone, but using both a cumulative distribution function and a probability density function with the mean clearly marked should result in the best comprehension of risk and uncertainty. Another obstacle to effective uncertainty communication is that the public may interpret a discussion of uncertainty as a sign of incompetence, or even dishonesty. Johnson and Slovic (1995) state that their results suggest "citizens find it hard to fathom that competence and uncertainty can coexist." A third obstacle to effective risk communication is that general risk attitudes or perceptions seem to be more influential than the presentation of uncertainty in the public's perception of risk (Johnson and Slovic 1995). Because of these obstacles, Johnson and Slovic (1995) advised "caution in assuming that explaining uncertainty will improve public trust or knowledge" and further stated "overall public trust and knowledge on risk issues may have to be built with methods more direct and difficult than uncertainty explanations."

VII. CONCLUSION

As noted in NRC (1994), uncertainty in risk assessment is a problem of significant proportion. Often, uncertainties arise where one must confront imperfect scientific knowledge or natural heterogeneity. The approaches historically used to address uncertainty have, unfortunately, been limited or misguided. The inadequate treatment of uncertainty has led to problems which the NRC (1983) hoped to avoid, such as the infusion of risk management decisions into the risk assessment process. As a result of inadequate treatment of uncertainty, past risk assessments have yielded conclusions that may be far from realistic and of limited scientific merit. More importantly, the inadequate treatment of uncertainty may have adversely impacted the measures to protect public health.

The solution to the problem lies in acknowledging the significance of uncertainties in risk assessment and better characterizing their sources and overall effects. A number of recent policies and guidelines have opened opportunities to characterize uncertainty adequately. The next major step is the implementation of these policies and guidelines into the regulatory framework. Presently, many tools and techniques are available to support quantitative characterization of uncertainty. It is foreseeable that with greater scientific and regulatory progress to address uncertainty, the risk assessment process will be more useful in addressing issues of environmental pollutants and their impacts on public health.

BIBLIOGRAPHY

American Industrial Health Council (AIHC). 1994. *Exposure Factors Sourcebook*, AIHC Environmental Health Risk Assessment Subcommittee, Exposure Factors Sourcebook Task Force, Washington, DC.

Baird, J.S., Cohen, J.T., et al. 1996. Noncancer risk assessment: Probabilistic characterization of population threshold doses, *Human and Ecological Risk Assessment* 2(1):79–102.

Barnes, D.G., Daston, G.P., et al. 1995. Benchmark dose workshop: Criteria for use of a benchmark dose to estimate a reference dose, *Regulatory Toxicology and Pharmacology* 21: 296–306.

Beck, B.D., Calabrese, E.J., et al. 1989. The use of toxicology in the regulatory process, In: *Principles and Methods of Toxicology, Second edition*, A. Wallace Hayes, Ed., Raven Press, Ltd.: New York.

Bogen, K.T. 1990. *Uncertainty in Environmental Health Risk Assessment*, Garland Publishing, Inc.: New York, NY.

Bogen, K.T. 1994. A note on compounded conservatism, *Risk Analysis* 14(4):379–381.

Bogen, K.T. 1995. Methods to approximate joint uncertainty and variability, *Risk Analysis* 15(3):411–419.

Bogen, K.T., Spear, R.C. 1987. Integrated uncertainty and interindividual variability in environmental risk assessment, *Risk Analysis* 7(4):427–436.

Burmaster, D.E., Anderson, P.D. 1994. Principles of good practice for the use of Monte Carlo techniques in human health and ecological risk assessments, *Risk Analysis* 14(4):477–481.

Calabrese, E.J. 1987. Animal extrapolation: A look inside the toxicologist's black box, *Environmental Science and Technology* 21(7):618–623.

Chankong, V., Haimes, Y.Y., et al. 1985. The carcinogenicity prediction and battery selection (CPBS) method: A Bayesian approach, *Mutation Research* 153(3):135–166.

Commission on Risk Assessment and Risk Management. 1997. Risk Assessment and Risk Management in Regulatory Decision-Making, Final Report.

Crouch, E.A.C. 1996. Uncertainty distributions for cancer potency factors: Laboratory animal carcinogenicity bioassays and interspecies extrapolation, *Human and Ecological Risk Assessment* 2:130–149.

Crump, K.S. 1984. A new method for determining allowable daily intakes, *Fundamental and Applied Toxicology* 4:854–871.

Dakins, M.E., Toll, J.E., et al. 1994. Risk-based environmental remediation: Decision framework and role of uncertainty, *Environmental Toxicology and Chemistry* 13(12):1907–1915.

Decisioneering Corp. 1990. Crystal Ball Software, Denver, CO.

Dourson, M. L., Stara, J.F. 1983. Regulatory history and experimental support of uncertainty (safety) factors, *Regulatory Toxicology and Pharmacology* 3:224–238.

Einstein, A. In: J. R. Newman, Ed., *The World of Mathematics*, Simon and Schuster: New York.

Environmental Protection Agency (EPA). 1986. Guidelines for Carcinogenic Risk Assessment, *51FR33992-34003*.

Environmental Protection Agency (EPA). 1989. *Risk Assessment Guidance for Superfund, Volume I.: Human Health Evaluation Manual (Part A)*, Publication No. 540/1-89/002, Office of Emergency and Remedial Response, U.S. Environmental Protection Agency, Washington, DC.

Environmental Protection Agency (EPA). 1992a. *Guidance on Risk Characterization for Risk Managers and Risk Assessors*, U.S. Environmental Protection Agency, Office of the Administator, Washington, DC.

Environmental Protection Agency (EPA). 1992b. A cross-species scaling factor for carcinogen risk assessment based on equivalence of mg/kg3/4/day, *57FR24152-24173*.

Environmental Protection Agency (EPA). 1992c. Guidelines for Exposure Assessment, Office of Research and Development, Office of Health and Environmental Assessment, Exposure Assessment Group, U.S. Environmental Protection Agency, Washington, DC, *57FR22888-22937*.

Environmental Protection Agency (EPA). 1992d. *Integrated Risk Information System (IRIS) Support Documentation*, Online, National Center for Environmental Assessment, U.S. Environmental Protection Agency, Washington, DC.

Environmental Protection Agency (EPA). 1994. Methods for Derivation of Inhalation Reference Concentrations and Application of Inhalation Dosimetry, Report No. EPA/600/8-90/066F, Office of Research and Development, U.S. Environmental Protection Agency, Washington, DC.

Environmental Protection Agency (EPA). 1996a. Proposed Guidelines for Carcinogen Risk Assessment, Report No. EPA/600/P-92/003C, Office of Research and Development, U.S. Environmental Protection Agency, Washington, DC.

Environmental Protection Agency (EPA). 1996b. *Exposure Factors Handbook*, Draft, Office of Research and Development, National Center for Environmental Assessment, U.S. Environmental Protection Agency, Washington, DC.

Environmental Protection Agency (EPA). 1997a. Dose–Response Modeling for 2,3,7,8-TCDD, Health Assessment Document for 2,3,7,8-Tetrachlorodibenzo-p-dioxin (TCDD) and Related Compounds, Report No. EPA/600/P-92/001C8, Office of Research and Development, U.S. Environmental Protection Agency, Washington, DC. January 1997 Workshop Review Draft.

Environmental Protection Agency (EPA). 1997b. *Policy for Use of Probabilistic Analysis in Risk Assessment at the U.S. Environmental Protection Agency*, Office of Research and Development, U.S. Environmental Protection Agency, Washington, DC, on-line.

Finley, B., Proctor, D., et al. 1994. Recommended distributions for exposure factors frequently used in health risk assessment, *Risk Analysis* 14(4):533–553.

Frey, H.C., Rhodes, D.S. 1996. Characterizing, simulating, and analyzing variability and uncertainty: An illustration of methods using an air toxics emissions example, *Human and Ecological Risk Assessment* 2(4):762–797.

Gratt, L.B. 1989. Uncertainty in air toxics risk assessment, for presentation at the 82nd Meeting and Exhibition of the Air and Waste Management Association, Anaheim, CA, June 25–30, 1989.

Grogan, P.J., Heinold, D.W., et al. 1988. Uncertainty in multipathway health risk assessments, for presentation at the 81st Annual Meeting of APCA, Dallas, TX, June 19–24, 1988.

Hattis, D., Anderson, E. 1993. What should be the implications of uncertainty, variability, and inherent biases/conservatism for risk management decision making? White paper presented at USEPA/University of Virginia Workshop: When and How Can You Specify a Probability Distribution When You Don't Know Much? Charlottesville, VA.

Hattis, D., White, P., et al. 1990. Uncertainties in pharmacokinetic modeling for perchloroethylene: I. Comparison of model structure, parameters, and predictions for low-dose metabolism rates for models derived by different authors, *Risk Analysis* 10:449–457.

Hattis, D., White, P., et al. 1993. Uncertainties in pharmacokinetic modeling for perchloroethylene: II. Comparison of model predictions with data for a variety of different parameters, *Risk Analysis* 13(6):599–610.

Hoffman, F.O., Hammonds, J.S. 1994. Propagation of uncertainty in risk assessments: The need to distinguish between uncertainty due to lack of knowledge and uncertainty due to variability, *Risk Analysis* 14(5):707–712.

Ibrekk, H., Morgan, M.G. 1987. Graphical communication of uncertain quantities to nontechnical people, *Risk Analysis* 7(4):519–529.

International Atomic Energy Agency (IAEA). 1989. Evaluating the reliability of predictions using environmental transfer models, Safety Practice Publications of the IAEA, *IAEA Safety Series* 100:1-106, STI/PUB/835, IAEA, Vienna, Austria.

Jayjock, M.A., Hawkins, N.C. 1993. A proposal for improving the role of exposure modeling in risk assessment, *American Industrial Hygiene Association Journal* 54(12):733–741.

Johnson, B.B., Slovic, P. 1995. Presenting uncertainty in health risk assessment: Initial studies of its effects on risk perception and trust, *Risk Analysis* 15(4):485–494.

Johnson, T., Capel, J. 1992. *A Monte Carlo Approach to Simulating Residential Occupancy Periods and its Application to the General U.S. Population,* U.S. Environmental Protection Agency, Office of Air Quality Planning and Standards, Emission Standards Division, Research Triangle Park, NC.

McKone, T.E. 1994. Uncertainty and variability in human exposures to soil contaminants through home-grown food: A Monte Carlo assessment, *Risk Analysis* 14(4):449–463.

McKone, T.E., Bogen, K.T. 1991. Predicting the uncertainties in risk assessment, *Environmental Science and Technology* 26(10):1674–1681.

Melnick, R.L., Kohn, M.C., et al. 1996. Implications for risk assessment of suggested non-genotoxic mechanisms of chemical carcinogenesis, *Environmental Health Perspectives* 104 (Suppl 1):123–134.

Morgan, M.G., Henrion, M. 1990. *Uncertainty: A Guide to Dealing with Uncertainty in Quantative Risk and Policy Analysis,* Cambridge University Press: Cambridge.

Munshi, U., Marlia, C. 1989. Role of uncertainty in risk assessment, for presentation at the 82nd Annual Meeting and Exhibition of AWMA, Anaheim, CA, June 25–30, 1989.

National Research Council. 1983. *Risk Assessment in the Federal Government: Managing the Process,* National Academy Press: Washington, DC.

National Research Council. 1991. *Human Exposure Assessment for Airborne Pollutants: Advances and Opportunities,* National Academy Press: Washington, DC.

National Research Council. 1994. *Science and Judgment in Risk Assessment,* National Academy Press: Washington, DC.

Palisade Corporation. 1994. @Risk Software, Newfield, NY.

Price, P.S., Sample. J., et al. 1992. Determination of less-than-lifetime exposures to point source emissions, *Risk Analysis* 12:367–382.

Price, P.S., Su, S.H., et al. 1996. Uncertainty and variation in indirect exposure assessments: An analysis of exposure to tetrachlorodibenzo-p-dioxin from a beef consumption pathway, *Risk Analysis* 16(2):263–277.

Swartout, J.C., Dourson, M.L., et al. 1994. An approach for developing probabilistic reference doses, Presentation given at the Annual Meeting of the Society for Risk Analysis, Baltimore, MD.

Thompson, K.M., Burmaster, D.E., et al. 1992. Monte Carlo techniques for quantitative uncertainty analysis in public health risk assessments, *Risk Analysis* 12(1):53–63.

Whitmyre, G.K., Driver J.H., et al. 1992. Human exposure assessment I: Understanding the uncertainties, *Toxicology and Industrial Health* 8(5):297–320.

Woodruff, T.J., Bois, F.Y., et al. 1992. Structure and parameterzation of pharmacokinetic models: Their impact on model predictions, *Risk Analysis* 12(1):189–201.

CHAPTER 8

Measurement of Indoor Air Contaminants

Lance A. Wallace

CONTENTS

1-56670-323-9/99/$0.00+$.50
© 1999 by CRC Press LLC

I. OVERVIEW

Over the last two to three decades, a great number of studies have been conducted on indoor air quality in buildings and residences. The following is a brief overview and history of some of the major studies. After this effort to set the stage, the principles of indoor air measurement will be discussed, using as examples the measurement methods employed in these important studies. Then, a more complete survey of these and other major studies and their findings will be provided. To interpret and extend these findings requires indoor air quality and exposure models; these models will be discussed to conclude the chapter.

II. INTRODUCTION

Early studies were conducted in the 1970s in northern Europe, particularly the Scandinavian countries (Berglund et al. 1982a; Mølhave and Møller 1979; Seifert and Abraham 1982; Skov and Valbjorn 1987; Skov et al. 1989, 1990). There are several reasons for the fact that indoor air pollution was first perceived to be a problem in this area at that time. The oil crisis of 1973 put a premium on saving energy, leading to new building practices that drastically reduced outdoor ventilation, thereby allowing indoor sources to build concentrations of pollutants to high levels. These building practices were first instituted in the northern countries, where they would have the greatest energy-saving effect. Also, Scandinavian countries use rather homogenous building practices, with many buildings of similar construction. This had the effect of making problems due to new construction practices immediately evident, as they simultaneously surfaced in many buildings. One of the first indoor air quality problems was noticed in approximately 100 Swedish preschools and was eventually traced to emissions of a compound (casein) from a self-leveling cement used in all the preschools. These early studies concentrated on volatile organic compounds (VOCs), using relatively inexpensive gas chromatographic methods followed by flame ionization detection (GC-FID). Monitors were placed in fixed locations indoors and outdoors.

One of the first and largest of U.S. studies was the Harvard Six-City Study (Spengler et al. 1980; Spengler et al. 1981; Spengler and Thurston 1983; Spengler et al. 1983; Dockery and Spengler 1981a; Dockery and Spengler 1981b) of particle exposures among several thousand children and adults. New fixed monitors were designed and employed in these studies and personal particle monitors were used for the first time in large-scale environmental studies. More than 15 years after the beginning of the Six-City Study, the authors published the finding that mortality had increased in these cities on days with higher particle levels (Dockery et al. 1992). This finding, repeated in other cities, resulted in estimates on the order of 20,000 to 50,000 deaths a year due to particle concentrations well below the current ambient air quality standards and, following one of the great environmental battles of the 1990s, led to the promulgation (62 FR 38652, July 18, 1997) of a new standard for fine (equal to or less than 2.5 microns in diameter) particulate matter.

In the early 1980s, the EPA Total Exposure Assessment Methodology (TEAM) study employed newly developed personal monitors to find that personal exposures to many toxic and carcinogenic volatile organic compounds (VOCs) were two to five times higher than outdoor concentrations, even though the outdoor concentrations were measured in heavily polluted areas such as northern New Jersey and Los Angeles (Wallace 1987; Wallace et al. 1982; Wallace et al. 1984; Wallace et al. 1985; Wallace et al. 1986). Subsequent studies in the Netherlands and Germany found similar results in those cities. All of these studies employed gas chromatography and mass spectrometry (GC-MS) to identify and quantify chemicals with much more certainty than previous methods (e.g., GC-FID). The apparent sources of many of these VOCs were consumer products (e.g., 1,1,1-trichloroethane and p-dichlorobenzene), building materials (e.g., toluene, xylene, and decane), personal activities such as smoking (e.g., benzene and styrene), driving (e.g., benzene, toluene, and xylenes), wearing drycleaned clothes (e.g., tetrachloroethylene), and even taking showers (e.g., chloroform from hot water).

In 1983-84, another EPA TEAM study of carbon monoxide (CO) took place in Denver, Colorado, and Washington, D.C. (Akland et al. 1985). More than 1,200 persons carried a CO monitor for a day to determine microenvironmental concentrations in vehicles, homes, churches, and schools. The most important sources of CO were found to be smoking, driving — particularly in heavy traffic, exposure to gas stoves, and exposure to environmental tobacco smoke (Ott et al. 1988; Wallace et al. 1988b). A continuous CO sensor was combined with a special data logger operated by the participant to record activities.

In the 1980s, several large-scale studies sponsored by the Gas Research Institute and the Southern California Gas Corporation determined that gas stoves and gas furnaces were important sources of nitrogen dioxide (NO_2) in homes (Colome et al. 1987; Ryan et al. 1988a; Ryan et al. 1988b; Ryan et al. 1992; Schwab et al. 1990). Kitchen levels were generally higher than living rooms which were higher than bedrooms, a gradient expected with greater distance from the source. Pilot lights were found to be responsible for about one-third of the total contribution of the gas stove to NO_2 levels. The replacement of the pilot light with electronic ignition in many homes is expected to reduce exposure from this source.

In 1985, a large-scale EPA TEAM study of pesticides in two cities (Jacksonville, Florida, and Springfield, Massachusetts) found that indoor concentrations of the thirty-two target pesticides were generally an order of magnitude higher than outdoor concentrations (EPA 1990c; EPA 1990d). Some of the highest-risk pesticides had been previously banned by the EPA, but were still present in air and in dust due to their long lives in soil. These were the chlorinated pesticides and termiticides DDT, DDE, aldrin, dieldrin, heptachlor, and chlordane. Personal monitors and analytical methods were newly developed or improved under EPA sponsorship for this study.

The first large-scale U.S. study attempting to link symptoms of office workers with environmental conditions in buildings occurred in 1989. In this study, 7,000 government workers at the U.S. Environmental Protection Agency (three buildings) and the Library of Congress (Madison Building) answered questionnaires about their symptoms while environmental variables were being measured simultaneously in their offices (NIOSH 1991a; NIOSH 1991b; NIOSH 1991c; EPA 1990a; EPA 1990b; EPA 1991). Although no single causal element was identified, a number of variables, including perceived dustiness, hot stuffy air, odors, and glare, were associated strongly (p<0.0001) with one or more symptom groups.

A fourth EPA TEAM study focused on inhalable particles, but also included analysis of metals, polyaromatic hydrocarbons (PAHs), and phthalate esters (Pellizzari et al. 1992; Özkaynak et al. 1996a; Özkaynak et al. 1996b; Clayton et al. 1993). The study indicated that two major indoor sources of particles in many homes in Riverside, California, were smoking and cooking, but that an unknown indoor source or sources accounted for even more particle emissions in most homes. For the first time, emissions profiles for certain elements were developed for smoking and cooking. Phthalate esters were found to be produced mainly indoors, but the specific sources could not generally be identified. No important indoor sources of PAHs were identified, but the mild fall season meant that no indoor combustion sources were in use. A subsequent wintertime study of a California community was more successful in identifying PAH sources, including wood stoves and cigarettes (Sheldon et al. 1993).

A series of studies on personal exposure to environmental tobacco smoke (ETS) using a number of newly developed measurement methods has been carried out in sixteen U.S. cities and in Leeds, England, Stockholm, Sweden, Turin, Italy, and Barcelona, Spain (Jenkins et al. 1996; Philips et al. 1994; Philips et al. 1996; Philips et al. 1997a; Philips et al. 1997b). About 100 nonsmoking persons in each of the U.S. cities, and between 190 and 225 persons in each European city, were recruited to carry a personal monitor for 24 hours, keep an activity diary, and supply saliva samples for cotinine determination. Using the Leeds study as an example of the results, persons with a smoking partner (N = 48) had mean exposures of 219 micrograms per cubic meter ($\mu g/m^3$), compared to about 170 $\mu g/m^3$ for persons without a smoking partner (N = 207), a difference of about 49 $\mu g/m^3$. Nicotine exposures averaged 4 $\mu g/m^3$ for those with a smoking partner and 1.3 $\mu g/m^3$ for those without.

The above studies are among the most important of those that have led to our present understanding of indoor air quality issues. It is a fact that in nearly every one of these studies, special measurement methods had only recently been developed, or had to be developed specifically for use in personal and indoor air environments. Thus, new measurement methods specifically aimed at indoor use have been essential to our understanding of indoor air quality issues. The following discussion highlights the most important and the most recent of these methods.

III. MEASUREMENT METHODS

A. Principles

Indoor air measurement methods operate under more strict requirements of size, bulkiness, airflow, and noise than their outdoor counterparts. High-volume samplers such as those used in most EPA outdoor monitoring sites cannot be used in a home or building, not only because of their bulkiness and noise, but also because their high airflows would completely change the air flow regime of the building. Because only low flows are possible indoors, only a small amount of material can be collected on filters. Thus, extremely stringent requirements must be placed on weighing the filters, including 24-hour or longer equilibration in weighing rooms that are controlled for temperature and humidity both before and after use in the sampler. The filters must be weighed to within about 10 millionths of a gram, and often collect only about 100 μg of material.

Measurement methods for personal exposure must meet even stricter requirements. Size must be reduced to about a liter, weight to less than a kilogram, and noise to a level low enough not to interfere with ordinary conversation. Airflow for pumps must also be reduced, such that filters may collect only 20 to 30 μg of inhalable or fine particles over a 12-hour period and must be weighed to within 5 μg to provide adequate precision. Sorbents may be limited to collecting 15 to 20 liters of air over a 12-hour period.

The remainder of this section is divided into three parts: particles, organic gases, and inorganic gases. The particles category includes particle-bound organics such as the heavier polyaromatic hydrocarbons (PAHs) and metals such as lead. The organic gases category includes volatile organics (VOCs) and semivolatile organics (SVOCs), such as PCBs, nitrosamines, and the lighter PAHs. The major inorganic gases of interest indoors include nitrogen oxides (NO_x), CO, and carbon dioxide (CO_2).

B. Particles

1. Sampling

Particles may be measured by determining their weight on a filter (i.e., gravimetric methods) or by counting them (i.e., optical methods). Historically, most of the measurement methods employed in environmental field studies have been gravimetric; there are no optical methods currently accepted by the EPA as reference methods. This may change as improved optical methods are developed and made widely available.

a. Gravimetric Particle Samplers

The traditional particle sampler basically consists of a pump to pull air at a known rate across a filter. The filter collects particles over a period of time and is then weighed. Although this sounds simple and straightforward, in fact it is often quite difficult to provide an accurate sample. Particularly for residential sampling, the greatest difficulty is the small amount of mass that can be collected and the resultant requirements for extreme care in handling such a small sample. Larger amounts can be collected by increasing either the flow rate or the sampling time. However, a very large pump flow rate indoors could result in changing the existing indoor conditions of velocity, airflow, and pressure, and the results would be meaningless. Therefore, as a practical matter it is not possible to sample at much above 20 liters per minute (Lpm). For a typical indoor PM_{10} (particulate matter equal to or less than 10 microns) concentration of 20 $\mu g/m^3$, an 8-hour sampling period would collect only about 200 μg of material. Considering that a typical filter weighs 100 mg, one is measuring the difference between 100 mg and 100.2 mg, a difference of only 0.2%. The filter can increase or decrease its weight by more than that due to absorbing water vapor due to changes in relative humidity. Because of this, filters must be conditioned for lengthy periods in humidity and temperature-controlled conditions both before and after sampling. Static charge can also cause spurious weight readings and, therefore, must be removed from the filters before weighing. Despite these problems, very careful field work can succeed in weighing filters reproducibly to within 10 μg. For the most recent EPA study of particles, the sampling period was 12 hours, the flow rate was 4 Lpm, and the average indoor concentration was 80 $\mu g/m^3$, leading to a total amount on the filter of only 230 μg on average, and much less than that in "clean" homes; yet the filters could be weighed to within 4 μg and the resulting precision was within 5%.

Various pump and filter samplers are available for personal and indoor sampling in residences. One type employs a 4 Lpm monitor for personal sampling and a 10 Lpm microenvironmental exposure monitor. The personal sampler collects one size fraction (either $PM_{2.5}$ or PM_{10}) on 37 mm filters (Marple et al. 1987). The indoor monitor can simultaneously collect both size fractions.

Methods for monitoring particles in ETS have been developed under the sponsorship of tobacco companies (Ogden et al. 1989; Ogden et al. 1990; Ogden et al. 1995). Since there are many sources of particles besides smoking, the thrust of these methods is toward finding a more specific tracer of ETS. One such tracer is UVPM, which refers to particles that respond to ultraviolet light, betraying a combustion origin. Since there are also other confounding combustion sources, a second tracer is FPM, particles that respond to a signal by fluorescing. Finally, an organic substance associated with ETS is solanesol, which can act as a tracer for the particle fraction of ETS, provided that the ratio of the solanesol to the total particle mass is known and invariant from one brand of cigarettes to another. The order of mass from nonspecific measures to measures specific for ETS would be expected to be RSP > UVPM > FPM $\approx$ solanesol. All three of these particle tracers have been used in recent large-scale studies of personal exposure to ETS in Europe and the U.S. However, in one of these studies, the expected order above was not observed, leading the authors to suggest that the solanesol–ETS particle mass relationship might be in need of further verification.

All such pump and filter samplers must be returned to the laboratory for weighing, no later than a day or two after the measurements. Therefore, they are unable to provide a real-time reading. One gravimetric instrument, the Piezobalance (manufactured by TSI, Minneapolis Minnesota), is able to provide a semi-real time reading. The Piezobalance uses an impactor to collect particles below 3.5 μm aerodynamic diameter at 1 Lpm. The particles pass through a chamber with an electrostatic precipitator that forces them to deposit on a piezoelectric crystal. The crystal is kept under forced vibration by an electric current. The vibration frequency changes as a result of the mass deposited (the piezoelectric effect). This change in frequency is monitored by continuous comparison with the frequency of an identical crystal in a closed chamber. The resulting cumulative change in frequency is divided by the time (usually two minutes, but the time can be lengthened under cases of very clean environments) to determine the total mass collected in that time. The frequency change is linear with mass up to a total mass of some 30 ng, at which time the monitor must be cleaned, a process that takes one to two minutes. Maintenance includes cleaning the needle assembly for the electrostatic precipitator and coating the impactor with an even layer of grease. The instrument is calibrated at the laboratory and is usually recalibrated once per year. The Piezobalance has never been tested under the EPA reference method program and so has failed to be selected as a reference sampler.

Nonetheless, the Piezobalance is a gravimetric sampler with the ability to make near real-time measurements and, therefore, has been used in a number of studies of indoor environments. One of the most well-known studies was carried out by Repace and Lowrey (1980) and resulted in documenting increased concentrations of RSP ($PM_{3.5}$) related to ETS in restaurants, offices, and other areas allowing smoking.

A more recent instrument is the TEOM (Tapered Element Oscillating Microbalance) sampler (Rupprecht et al. 1992). This instrument uses a tapered oscillating rod whose oscillation frequency is altered by the mass of particles settling on it. The instrument has been used effectively for outdoor monitoring and has found occasional use as an indoor monitor. However, because the element must be heated to about 40–50°C, it is thought that particle-bound organics, which may account for up to half of particle mass, may be driven off on encountering the high temperatures of the element (Koutrakis et al. 1992)

b. Automated Particle Counters

Recently, particle counters capable of simultaneously measuring six or eight size ranges have become commercially available. One type uses laser diode technology to provide counts of particles in six size ranges, from 0.3 μm to 10 μm. A built-in printer prints out the data at programmable intervals. A version of this monitor uses a manifold allowing automatic sampling from several locations. Portable (hand-held) versions are also available for personal sampling. Count data are stored in memory for transfer to a computer.

Far more complex particle counters capable of providing data on very finely divided size fractions are beginning to find use in field monitoring indoors. One such monitor is the Aerodynamic Particle Sizer (APS) of TSI, Inc., which returns information on particle numbers in 50 size fractions from about 0.1 μm to about 20 μm, although the range from 0.1 μm to 1 μm is of uncertain validity. A complementary system, returning data on the ultrafine particles from about 0.01 μm to 0.5 μm, is the Scanning Mobility Particle Sizer (SMPS) by the same company. These two devices have been used in research studies of a few residences (Abt Inc., personal communication).

Although these monitors provide useful data on numbers of particles, they can be misleading when used to estimate mass because of widely varying particle densities. If the aerosol is of known composition (e.g., ETS), the monitors can be calibrated to the particular aerosol and provide trustworthy mass information. But, if the aerosol is of unknown composition it is presently impossible to obtain accurate mass information from optical counters.

2. Analysis

a. Elemental analysis

Particles collected on filters can be analyzed for their elemental content using either X-ray fluorescence (XRF) or proton-induced X-ray emission (PIXE). Teflon filters are often employed to avoid the high elemental background of quartz fiber filters. Depending on air concentrations and volume of air sampled, up to 30 or 40 elements can be analyzed using one of these methods. In the PTEAM Study, which used monitors collecting only 3 m³ of air, 14 elements were commonly detectable by XRF. Besides providing data directly on toxic metals like lead and cadmium, metal analysis can also be used for source apportionment (e.g., marker elements such as silicon for crustal material and vanadium for home heating oil).

b. Organic Analysis

Particle-bound organics such as the heavier PAHs can also be analyzed following collection of a sample. Since the lighter PAHs are often in the vapor phase, and medium-weight PAHs may exist in both aerosol and gaseous states, a combination of a filter and a sorbent is sometimes used to collect all the PAHs in both states, with subsequent extraction of the filter and sorbent together. This method provides an accurate total but does not allow for identification of the relative amounts in the vapor and particle stage. A method that can determine the phase distribution more exactly employs a diffusion denuder followed by a filter and sorbent (Coutant et al. 1985; Coutant et al. 1986).

A continuous monitor for total PAHs has also been developed and tested in the field (Wilson et al. 1995b). By collecting indoor and outdoor data simultaneously, the impact of woodsmoke, for example, on indoor air quality can be determined. Methods for analyzing the many constituents of tobacco smoke are provided in several sources (Guerin et al. 1992; Ogden et al. 1989; Ogden et al. 1990; Ogden et al. 1995; Löfroth et al. 1989).

C. VOCs

1. Sampling Methods

a. Activated Charcoal

An early method developed for occupational exposures (generally for concentrations in the range of 10–100 parts per million for a given VOC) was the use of a sorbent (usually activated charcoal) in order to concentrate the VOCs. They are then recovered by a solvent such as carbon disulfide and analyzed by gas chromatography.

In the early 1980s, passive badges employing activated charcoal were developed for use in occupational sampling. The badges operate on the principle of diffusion and often are operated over an 8-hour workday to provide an integrated average exposure for comparison to the occupational standards (e.g., the threshold limit value, or TLV). The manufacturing process for these badges leaves residues of VOCs on the activated carbon making the badges unsuitable for short-term sampling at environmental concentrations, which are usually at part-per-billion (ppb) levels. However, the high background contamination on the badges can be overcome by extending the time of sampling to a week or more, and several studies of indoor air pollution have adopted this technique (Mailahn et al. 1987; Seifert and Abraham 1983).

b. Tenax

The background problems associated with activated charcoal, as well as problems in obtaining reliable recoveries of sorbed chemicals, led to a search for a more suitable sorbent. A polymer known as Tenax was widely adopted during the 1970s as a more reliable sorbent than charcoal for ppb levels (Barkley et al. 1980; Krost et al. 1982). Tenax, properly cleaned, has very low background contamination for almost all VOCs of interest. It also is stable at temperatures up to 250°C, allowing thermal desorption instead of solvent desorption.[1] Drawbacks include artifact formation of several chemicals (e.g., benzaldehyde and phenol) and an inability to retain highly volatile organic chemicals (e.g., vinyl chloride and methylene chloride). Although most uses of Tenax sorbent have been with active (pumped) samplers, a passive badge containing Tenax has also been developed (Coutant et al. 1985; Coutant et al. 1986; Lewis et al., 1985).

c. Multisorbent Systems

In the late 1980s, attempts were made to combine the best attributes of charcoal and Tenax into a multisorbent system. Newer types of activated charcoal (e.g., Spherocarb and Carbosieve) were developed to provide more reliable recoveries. Tandem systems employing Tenax as the first sorbent and activated charcoal as the second, or backup, sorbent were employed. The Tenax collected the bulk of the VOCs and the activated charcoal collected those more volatile VOCs that "broke through" the Tenax. Systems were also developed using three sorbents, such as Tenax, Ambersorb, and Spherocarb or Carbosieve (Hodgson et al. 1986). All such systems allow collection of a broader range of chemical types and volatilities.

d. Direct (Whole Air) Sampling

This method, first developed in the 1970s for upper atmosphere sampling, avoids the sorption-desorption step, which should theoretically allow less chance for contamination; however, it requires great sensitivity on the part of the detection instruments. The method may involve real-time sampling in mobile laboratories, with

---

[1] Solvent desorption involves a redilution of the VOCs, thus partially negating the concentration made possible by the sorbent.

direct injection of the air sample into a cold trap attached to a GC, or sampling in evacuated electropolished aluminum canisters for later laboratory analysis (Oliver et al. 1986).

e. Comparison of Sampling Methods

No single method of sampling VOCs in the atmosphere or indoors has become a standard or reference method. In the U.S., the two preferred methods are Tenax and evacuated canisters. These two methods were compared under controlled conditions in an unoccupied house (Spicer et al. 1986). Ten chemicals were injected at nominal levels of about 3, 9, and 27 $\mu g/m^3$. The results showed that the two methods were in excellent agreement, each with a precision of better than 10% for all chemicals at all spiked levels.

In Europe, the two most common methods are Tenax and activated charcoal. One study employing both methods side by side (Skov et al. 1990) found consistently higher levels of total VOC on the charcoal sorbent. The difference may be due to very volatile organics such as pentane and isopentane, which are collected by charcoal but which break through Tenax readily.

The sorbent methods lend themselves to personal monitoring—a small battery-powered pump is worn for an 8-hour or 12-hour period to provide a time-integrated sample. At present, however, the whole-air methods employ bags or canisters that are too bulky or heavy to be used as personal monitors.

2. Analysis

Samples are usually analyzed by first separating the components using gas chromatography (GC). Three detection methods in common use are flame ionization (FID), electron capture (ECD), and mass spectrometry (MS). Only GC-MS has the ability to identify unambiguously many chemicals. Neither GC-FID nor GC-ECD is able to separate chemicals that coelute (i.e., emerge from the chromatographic column at the same time). Also, GC-FID response is depressed by halogens such as chlorine, so it is not suitable for samples containing halogens. On the other hand, GC-ECD is extremely well suited to measuring halogens at very low concentrations. Mass spectrometry, by breaking chemicals into fragments and then identifying these fragments, often is capable of differentiating even among coeluting chemicals. However, since chemicals are identified by comparing these mass fragment spectra to existing libraries, and the libraries are incomplete, even GC-MS identifications are often tentative or mistaken. For example, one study using known mixtures of chemicals found about 75% accuracy of identification for several different GC-MS computerized spectral search systems.

D. Pesticides

Work sponsored by the EPA has succeeded in developing indoor and personal monitors for pesticides (Lewis and MacLeod 1982). The basic sorbent employed is polyurethane foam (PUF), which can be analyzed for scores of the "traditional" chlorinated pesticides, including DDT, chlordane, aldrin, dieldrin, and others in this general category. Some of the newer organophosphate pesticides such as chlorpyrifos can also be collected on PUF and analyzed by multianalyte methods.

Although these personal and indoor monitors have been used in field studies, none are commercially available. Also, analysis continues to be extremely expensive, on the order of $1,000 per sample to identify a suite of common pesticides. Cost savings can be achieved if only one or a few pesticides are the targets, but analyses of pesticides are generally too difficult and expensive to allow for much data to be collected in homes.

E. Nitrosamines

Recently, a convenient sampling and analysis method was developed for determining volatile N-nitrosamines in ETS, of interest because of their potency as carcinogens (Mahanama and Daisey 1996). The method employs commercially available Thermosorb/N cartridges to collect and fix the N-nitrosamines. Excellent recoveries of 96 ± 5% were obtained for the four most common N-nitrosamines. The authors modified the manufacturer's recommendation for an extraction-cleanup method to produce an increase in sensitivity to about 17–23 ng/cigarette for the three most common N-nitrosamines found in ETS.

F. Carbon Monoxide

A large number of continuous monitors, both active (pumped) and passive (diffusion), are commercially available for CO. Most are electrochemical, depending on counting the electrons produced when CO is oxidized to CO_2. Precision of these monitors is generally very good, with typical errors on the order of 0.1 ppm. Interferences can be a problem in some cases. For example, when the monitors are used for breath analysis, endogenously produced hydrogen (from eating certain foodstuffs such as beans) can be a positive interference.

G. Nitrogen Oxides

A useful monitor for NO_2 is the Palmes Tube (Palmes et al. 1986), which consists of a short plastic tube with a filter soaked with a solution of triethanolamine, which reacts with NO_2 and can later be quantitated by a colorimetric method. The system has a sensitivity of about 600 ppb/h, so that for typical environmental concentrations of 10–20 ppb, a sampling period of a few days is sufficient to obtain a measurement.

An improvement in some ways over the Palmes Tube is the Yanagisawa Badge (Yanagisawa and Nishimura 1982), which employs a baffle to reduce sensitivity to wind velocity, and a smaller length-to-area distance to improve diffusion rate; the badge is about ten times as sensitive as the Palmes Tube, allowing shorter collection periods.

IV. AIR EXCHANGE

Air exchange is one of several crucial ancillary variables in understanding indoor-outdoor relationships. Air exchange rates can have an important influence on pollution levels in the home. For example, if outdoor air is cleaner than indoor air, as is true for most VOCs and most pesticides (as will be shown below), then it will improve matters to open the windows, turn on the attic fan, or otherwise increase the air exchange rate. On the other hand, during times of high outdoor air pollution, closing the windows or otherwise reducing air exchange rates can have a protective effect. This is particularly true for particles, which deposit on walls and thus are removed from indoor air—the lower the air exchange rate, the more particles will be removed from the air. Tracer gases such as sulfur hexafluoride (SF_6) or other perfluorinated tracers (PFTs) not found in nature are generally used to measure air exchange rates (Dietz and Cote 1982).

In one approach, a source with a constant emission rate is placed in the home for the duration of the monitoring period. A collector device, often a sorbent such as activated charcoal, collects an amount of tracer inversely proportional to the total outdoor air flow through the house. A second approach also uses a continuous source but utilizes a continuous direct readout device to provide a more detailed temporal record of air exchange. A third approach is to introduce the tracer into the home over a limited time period and then to record its decay. The latter approach is generally good only for shorter time intervals of 8 to 16 hours, but has the advantage of providing a direct indication of the air exchange rate without having to determine the volume of the structure.

In the first two methods, division of the air flow rate (m^3/h) by the volume of the house (m^3) gives the air exchange rate, in inverse hours. That is, an air exchange rate of 1 h^{-1} indicates that the home exchanges its volume of air with the outdoors every hour. In fact, under a well-mixed scenario, a fraction ($1/e$) of the air molecules in the home at the beginning of the hour remain at the end of the hour.

Use of the perfluorinated tracers provides the most accurate measurement of air exchange, but less expensive methods can also be employed. Because of the general decline of CO emissions from automobiles, many homes in suburban neighborhoods have very low CO concentrations just outside; for such homes with gas stoves, the decay of CO concentrations in the home just after extensive cooking has been used to measure air exchange rates of the home.

V. MAJOR INDOOR AIR STUDIES

A. Concentrations and Sources

1. Particles

Four large-scale studies of airborne particles have been performed that present the most complete investigations to date of indoor and outdoor concentrations of particles. In chronological order, these are:

1. The Harvard Six-City Study, sponsored by the National Institute for Environmental Health Sciences (NIEHS) and the EPA and carried out by the Harvard School of Public Health beginning in 1979 and continuing through 1988, with measurements taken in at least 1,400 homes (Spengler et al. 1981);
2. The New York State study, sponsored by the New York State Energy Research and Development Authority (NYSERDA), and carried out by Research Triangle Institute in 433 homes in two New York State counties in 1986 (Sheldon et al. 1989);
3. The EPA Particle TEAM (PTEAM) Study, carried out by Research Triangle Institute and Harvard University School of Public Health in 178 homes in Riverside, California in 1990 (Pellizzari et al. 1992);
4. ETS studies in sixteen U.S. cities (about 1,500 participants) and a number of major European cities (about 200 participants per city) sponsored by the Center for Indoor Air Research (CIAR) (Jenkins et al. 1995a; Jenkins et al. 1995b).

a. Harvard Six-City Study

The Harvard Six-City Study was a prospective epidemiological study of the effects of particles and sulfur oxides on the health of grade school children. The six cities were chosen to represent low (Portage, Wisconsin and Topeka, Kansas), medium (Watertown, Massachusetts and Kingston-Harriman, Tennessee), and high (St. Louis, Missouri and Steubenville, Ohio) outdoor particle and sulfate levels.

The study took place in two measurement phases. The first phase involved monitoring about ten homes in each city for respirable particles ($PM_{3.5}$). The homes were measured every sixth day (24-hour samples) for one to two years. In the second phase, a larger sample of 200 to 300 homes was selected from each city, with two consecutive week-long $PM_{2.5}$ samples collected both indoors and outdoors in summer and again in winter. More than 1,400 homes were monitored.

Spengler et al. (1981) described the first five years of the Harvard Six-City Study. During that time pulmonary function measurements were administered to 9,000 adults and 11,000 children in grades 1 to 6. Mean indoor concentrations exceeded the outdoor levels in all cities except Steubenville, where the outdoor levels of about 46 $\mu g/m^3$ slightly exceeded the indoor mean of about 43 $\mu g/m^3$. The authors noted that the major source of indoor particles is cigarette smoke, with levels of 24 $\mu g/m^3$ in homes without smokers, 36 $\mu g/m^3$ in homes with one smoker, and 70 $\mu g/m^3$ in homes with two smokers.

Neas et al. (1994) presented summary results for the entire second phase of the Six-City Study (1983–88). Homes with children 7 to 11 years old whose parents had never smoked were eligible for the study. A total of 1,273 homes with children completed two weeks of summer and winter monitoring indoors and outdoors for $PM_{2.5}$ using the Harvard impactor. The annual (winter and summer) household $PM_{2.5}$ mean concentration for the 580 children living in consistently smoking households was 48.5 ± 1.4 $\mu g/m^3$ compared to 17.3 ± 0.5 $\mu g/m^3$ for the 470 children in consistently nonsmoking households.

Spengler et al. (1987) reported on a new round of measurements in three communities—Watertown, Massachusetts, St. Louis, Missouri, and Kingston-Harriman, Tennessee—within the Six-City Study. In each community, about 300 children were selected to take part in a year-long diary and indoor air quality study. Measurements of PM$_{2.5}$ were taken indoors at home for two consecutive weeks in winter and again in summer. The sampler was the automated Harvard sampler (Marple et al. 1987), which collected an integrated sample for the week except for the five 8 A.M. to 4 P.M. weekday periods when the child was at school. During this 40-hour period, samples were taken in one classroom in each of the elementary schools involved. Results were presented for smoking and nonsmoking homes in each city by season. The authors noted that mean concentrations in homes with smokers were about 30 µg/m^3 greater than homes without smokers. The difference was greater in winter than in summer in all cities.

Santanam et al. (1990) reported on a more recent and larger-scale monitoring effort in Steubenville and Portage as part of the Six-City study. In each city, 140 homes, split evenly between those with and without smokers, were monitored for one week in summer and in winter. The Harvard impactor sampler was used with an automatic time unit to collect PM$_{2.5}$ samples between 4 P.M. and 8 A.M. on weekdays and all day on weekends, corresponding to likely times of occupancy for school-age children. Outdoor samples were collected from one site in each city. Elements were determined by X-ray fluorescence (XRF). A source apportionment using principal components analysis (PCA) and linear regressions on the elemental data was performed. Cigarette smoking was the single largest source, accounting for 20–27 µg/m^3 indoor PM$_{2.5}$ in Steubenville and 10–25 µg/m^3 in Portage. Homes with smokers exceeded outdoor levels by 25 (summer) and 20 (winter) µg/m^3 in Steubenville, and 24 and 11 µg/m^3 in Portage. Neas at al. (1994) carried out additional analyses on exposures of children, finding a strong effect of ETS on increased exposures (See Figure 8.1).

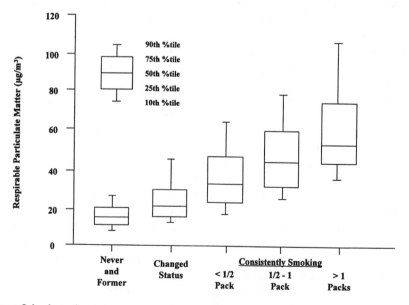

Figure 8.1 Annual average concentrations of indoor PM$_{25}$ by household smoking status and estimated number of cigarettes smoked in the home.

b. The New York State Study

Sheldon et al. (1989) studied $PM_{2.5}$ and other agents in 433 homes in two New York State counties. One goal of the study was to determine the effect of kerosene heaters, gas stoves, wood stoves or fireplaces, and cigarette smoking on indoor concentrations of combustion products. A stratified design to include all 16 combinations of the four combustion sources was implemented, requiring about 22,000 telephone calls.

The sampler was a portable dual-nozzle impactor (Marple et al. 1987). Samples were collected in the main living area and in one other room (containing a combustion source if possible) during alternate 15-minute periods over a 7-day period. Outdoor samples were collected at a subset of 57 homes. All samples were collected during the winter (January–April) of 1986.

Mean indoor $PM_{2.5}$ concentrations were approximately double those outdoors in both counties. Of the four combustion sources, only smoking created significantly higher indoor $PM_{2.5}$ concentrations in both counties. Use of kerosene heaters was associated with significantly higher concentrations in Suffolk (N = 22) but not in Onondaga (N = 13). Use of wood stoves/fireplaces and gas stoves did not elevate indoor concentrations in either county.

Leaderer and Hammond (1991) continued their analysis of the New York State data by selecting a subset of 96 homes for which both nicotine and $PM_{2.5}$ data were obtained. In the 47 homes in which nicotine was detected (detection limit = 0.1 $\mu g/m^3$), the mean concentration of RSP was 44.1 (± 25.9 SD) $\mu g/m^3$ compared to 15.2 (± 7.4) $\mu g/m^3$ in the 49 homes where no nicotine was detected. Thus, homes with smoking had an increased weekly geometric mean $PM_{2.5}$ concentration of about 29 $\mu g/m^3$. Regressions of $PM_{2.5}$ on total number of cigarettes smoked during the week (N_{cig}) gave the results:

$$PM_{2.5} = 17.7 + 0.322 \ N_{cig} \qquad (N = 96; \ R^2 = 0.55)$$

$$PM_{2.5} = 24.8 + 0.272 \ N_{cig} \qquad (N = 47; \ R^2 = 0.40)$$

where the first regression includes all homes but the second includes only homes with measurable nicotine levels. Thus, each cigarette produces an increase of about 0.3 (± 0.03) $\mu g/m^3$ in the weekly mean $PM_{2.5}$ concentration, equivalent to an increase of 2.1 (± 0.2) $\mu g/m^3$ in the daily concentration.

Koutrakis et al. (1992) also analyzed the New York State data, using a mass-balance model to estimate $PM_{2.5}$ and elemental source strengths for cigarettes, wood burning stoves, and kerosene heaters. $PM_{2.5}$ source strength for smoking was estimated at 12.7 $\pm$ 0.8 (SE) mg/cigarette. For a final category of all other residual indoor sources, a source strength of 1.16 mg/h was calculated. For nonsource homes (N = 49) the authors estimated that 60% (9 $\mu g/m^3$) of the total $PM_{2.5}$ mass was from outdoor sources, and 40% (6 $\mu g/m^3$) from unidentified indoor sources. For smoking homes, they estimated that 54% (26 $\mu g/m^3$) of the $PM_{2.5}$ mass was from smoking, 30% (15 $\mu g/m^3$) from outdoor sources, and 16% (8 $\mu g/m^3$) from unidentified sources.

These authors also developed an elemental emissions profile for $PM_{2.5}$ particles from cigarettes, woodburning, and kerosene heaters. For cigarettes, the elemental profile included potassium (160 µg/cig), chlorine (69 µg/cig), and sulfur (65 µg/cig), as well as smaller amounts of bromine, cadmium, vanadium, and zinc. The woodburning profile included three elements: potassium (92 µg/h), silicon (44 µg/h), and calcium (38 µg/h). The kerosene heater profile included a major contribution from sulfur (1,500 µg/h) and fairly large inputs of silicon (195 µg/h) and potassium (164 µg/h). A drawback of the mass-balance model was an inability to estimate separately the value of the penetration coefficient P and the decay rate k for particles and elements; Koutrakis et al. (1992) assumed a constant rate for k of 0.36 h^{-1}, and then solved for P.

c. The EPA Particle TEAM (PTEAM) Study

In 1986, the U. S. Congress mandated that the EPA Office of Research and Development carry out a TEAM Study of human exposure to particles. The EPA Atmospheric Research and Exposure Assessment Laboratory (AREAL) joined with the California Air Resources Board to sponsor a study in the Los Angeles Basin. The study was carried out primarily by the Research Triangle Institute and the Harvard School of Public Health, with additional support from Lawrence Berkeley Laboratory, Acurex, and AREAL. The main goal of the study was to estimate the frequency distribution of exposures to particles for Riverside residents. Another goal was to determine particle concentrations in the participants' homes and immediately outside the homes.

A pilot study was undertaken in nine homes in Azusa, California, in March of 1989 to test the sampling equipment. Newly designed personal exposure monitors (PEMs) were equipped with inhalable (PM_{10}) and fine ($PM_{2.5}$) particle inlets. The PEMs were impactors with 4 Lpm Casella pumps. Two persons in each household wore the PEMs for two consecutive 12-hour periods (night and day). Each day they alternated inlet nozzles. The first five households were monitored concurrently for seven consecutive days; the last four households were then monitored concurrently for four consecutive days. This resulted in approximately 100 PEM samples for each size fraction. Indoor and outdoor particle concentrations were monitored using microenvironmental exposure monitors (MEMs). These monitors were the Harvard "black boxes" (Marple et al., 1987) employing a 10 Lpm pump. Several indoor MEMs were placed in different rooms in each house to determine the magnitude of room-to-room variation. These monitors were capable of simultaneously measuring both fine and inhalable particles. Room-to-room variation of 12-hour average particle levels was generally less than 10%. It was decided that this finding would justify using only one indoor monitor in the subsequent full-scale study.

The personal exposures were about twice as great as the indoor or outdoor concentrations for both PM_{10} and $PM_{2.5}$. Considerable effort was expended to determine whether this was a sampling artifact due, for example, to the constant motion of the personal sampler; however, no evidence could be found for an artifactual effect. Nonetheless, to reduce chances for an artifactual finding in the main study, it was decided to use identical PEMs for both the personal and fixed (indoor-outdoor) samples in the main study.

Regressions of outdoor on indoor concentrations showed low R^2 values (1–30%) for both PM_{10} and $PM_{2.5}$ size fractions, as did regressions of daytime indoor on personal concentrations ($R^2 = 0$–18%). Overnight indoor concentrations had somewhat better ability to explain personal exposures ($R^2 = 14$–58%), as might be expected from the fact that the personal monitor was placed on the bedside table during the sleeping period. Personal exposures were essentially uncorrelated with outdoor concentrations ($R^2 = 0$–2%).

For the main study the following year, a three-stage probability sampling procedure was adopted (Pellizzari et al. 1992). Ultimately, 178 residents of Riverside, California, took part in the study in the fall of 1990. Respondents represented $139,000 \pm 16,000$ (SE) nonsmoking Riverside residents aged ten and above. Their homes represented about 60,000 Riverside homes.

Each participant wore the PEM for two consecutive 12-hour periods. Concurrent PM_{10} and $PM_{2.5}$ samples were collected by the stationary indoor monitor (SIM) and stationary outdoor monitor (SAM) at each home. The SIM and SAM were essentially identical to the PEM. A total of ten particle samples were collected for each household (day and night samples from the PEM_{10}, SIM_{10}, $SIM_{2.5}$, SAM_{10}, and $SAM_{2.5}$). Air exchange rates were also determined for each 12-hour period.

Up to four participants per day could be monitored, requiring 48 days in the field. A central outdoor site was maintained over the entire period (Sept. 22 to Nov. 9, 1990). The site had two high-volume samplers (Wedding and Assoc.) with 10-μm inlets (actual cutpoint about 9.0 μm), two dichotomous PM_{10} and $PM_{2.5}$ samplers (Sierra-Andersen) (actual cutpoint about 9.5 μm), one PEM, one PM_{10} SAM, and one $PM_{2.5}$ SAM.

More than 2,750 particle samples were collected, about 96% of those attempted. All filters were analyzed by X-ray fluorescence (XRF) for a suite of 40 metals. More than 1,000 12-hour average air exchange rate measurements were taken.

A complete discussion of the quality of the data is found in Pellizzari et al. (1992) and in Thomas et al. (1993a). Limits of detection (LODs), based on three times the standard deviation of the blanks, were on the order of 7–10 μg/m³. All field samples exceeded the LOD. Duplicate samples (N = 363) showed excellent precision for all types of particle samplers at all locations, with median relative standard deviations ranging from 2–4%.

Daytime mean personal PM_{10} concentrations (150 μg/m³) were more than 50% higher than either indoor or outdoor levels (95 μg/m³). Overnight mean personal PM_{10} concentrations (77 μg/m³) were similar to the indoor (63 g/m³) and outdoor (86 μg/m³) levels. About 25% of the population of Riverside was estimated to have 24-hour personal PM_{10} exposures exceeding the 150 μg/m³ 24-hour national ambient air quality standard (NAAQS) for ambient air. The higher personal concentrations do not appear to be due to skin flakes or clothing fibers; many skin flakes were found on filters (up to an estimated 150,000 per filter) in subsequent scanning electron microscope (SEM) analyses, but their mass does not appear to account for more than 10% of the excess personal exposure.

Mean PM$_{2.5}$ daytime concentrations were similar indoors (48 µg/m³) and outdoors (49 µg/m³), but indoor concentrations fell off during the sleeping period (36
µg/m³) compared to 50 µg/m³ outdoors. Thus, the fine particle contribution to PM$_{10}$
concentrations averaged about 51% during the day and 58% at night both indoors
and outdoors.

Unweighted distributions are shown in Figure 8.2 for 24-hour average PM$_{10}$
personal, indoor, and outdoor concentrations. About 25% of the population of Riverside was estimated to have 24-hour personal PM$_{10}$ exposures exceeding the 150
µg/m³ 24-hour NAAQS for ambient air.

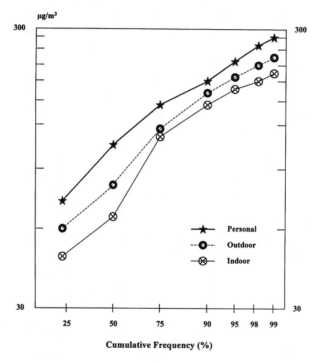

Figure 8.2   Particles in Riverside — 24-hour PM$_{10}$ concentration.

Central-site PM$_{2.5}$ and PM$_{10}$ concentrations agreed well with backyard concentrations. Overall, the data strongly suggest that a single central-site monitor can
represent well the PM$_{2.5}$ and PM$_{10}$ concentrations throughout a wider area such as
a town or small city, at least in the Los Angeles basin.

Stepwise regressions resulted in smoking, cooking, and either air exchange rates
or house volumes being added to outdoor concentrations as significant variables.
Smoking added about 30 µg/m³ to the total PM$_{2.5}$ concentrations. Cooking added 13
µg/m³ to the daytime PM$_{2.5}$ concentration, but was not significant during the overnight
period. At night, an increase in air exchange of one air change per hour resulted in
a small increase of about 4.5 µg/m³ to the PM$_{2.5}$ concentration, but was not significant
during the day. The house volume was not significant at night, but was significant

during the day, with larger homes resulting in smaller $PM_{2.5}$ concentrations. Since air exchange and house volume were weakly correlated (negatively), they were not included together in the same regression.

A model developed in Koutrakis et al. (1992) was solved using nonlinear least squares to estimate penetration factors, decay rates, and source strengths for particles and elements from both size fractions. Penetration factors are very close to unity for nearly all particles and elements. The calculated decay rate for fine particles is 0.39 ± 0.16 h^{-1}, and for PM_{10} is 0.65 ± 0.28 h^{-1}. Since PM_{10} contains the $PM_{2.5}$ fraction, a separate calculation was made for the coarse particles ($PM_{10} - PM_{2.5}$) with a resulting decay rate of 1.01 ± 0.43 h^{-1}. Each cigarette emits 22 ± 8 mg of PM_{10} on average, about two thirds of which (14 ± 4 mg) is in the fine fraction. Cooking emits 4.1 ± 1.6 mg/min of inhalable (PM_{10}) particles, of which about 40% (1.7 ± 0.6 mg/min) is in the fine fraction. All elements emitted by cooking were limited almost completely to the coarse fraction; presumably carbon or other elements not measured by XRF were contained in the fine fraction. Sources other than cooking and smoking emit about 5.6 ± 3.1 mg/h of PM_{10}, of which only about 1.1 mg/h ± 1.0 (20%) is in the fine fraction.

Decay rates for elements associated with the fine fraction were generally lower than for elements associated with the coarse fraction, as would be expected, due to their lower settling velocities. For example, sulfur, which has the lowest mass median diameter of all the elements, had calculated decay rates of 0.16 ± 0.04 and 0.21 ± 0.04 h^{-1} for the $PM_{2.5}$ and PM_{10} fractions, respectively. The crustal elements (Ca, Al, Mn, Fe), on the other hand, had decay rates ranging from 0.6 to 0.8 h^{-1}.

Based on the mass-balance model, outdoor air was the major source of indoor particles in Riverside, providing about 75% of fine particles and 65% of inhalable particles in the typical home. It was also the major source for most elements, providing 70 to 100% of the observed indoor concentrations for 12 of the 15 elements. Only the presence of copper and chlorine was predominantly due to indoor sources in both the fine particle and inhalable particle fractions. However, these conclusions are strictly applicable only to Riverside. In five of the six cities studied by Harvard, and in both New York counties, outdoor air could not have provided as much as half of the indoor air particles, since the observed indoor-outdoor ratios of mean concentrations were greater than or equal to two.

Unidentified indoor sources accounted for most of the remaining particle and elemental mass collected on the indoor monitors. The nature of these sources is not yet understood. They do not include smoking, other combustion sources, cooking, dusting, vacuuming, spraying, or cleaning, since all these sources together account for less than the unidentified sources. For example, the unidentified sources accounted for 26% of the average indoor PM_{10} particles, whereas smoking accounted for 4% and cooking for 5%.

Of the identified indoor sources, the two most important were smoking and cooking. Smoking was estimated to increase 12-hour average indoor concentrations of PM_{10} and $PM_{2.5}$ by 2 and 1.5 $\mu g/m^3$ per cigarette, respectively. Homes with smokers had mean PM_{10} concentrations about 30 $\mu g/m^3$ higher than homes without smokers. Most of this increase was in the fine fraction. Cooking increased indoor concentrations of PM_{10} by about 6 $\mu g/m^3$ per minute of cooking, with most of the increase in the coarse particles.

Emission profiles for elements were obtained for smoking and for cooking. Major elements emitted by cigarettes were potassium, chlorine, and calcium. Elements associated with cooking included aluminum, iron, calcium, and chlorine.

Multivariate calculations in all three studies result in rather similar estimates of the effect of smoking on indoor fine particle concentrations. Spengler et al. (1981) estimated an increase of about 20 $\mu g/m^3$ per smoker, or 25 $\mu g/m^3$ per smoking home, based on 55 smokers monitored over a year in all six cities. Spengler et al. (1985) found a smoking effect of about 32 $\mu g/m^3$ for smoking homes in multivariate models based on the Kingston-Harriman data. Sheldon et al. (1989) found an increase of 45 (Onondaga) and 47 (Suffolk) $\mu g/m^3$ per smoking home in a multivariate model of the New York State data. Özkaynak et al. (1993) found an increase of about 30–35 $\mu g/m^3$ in smoking homes in a multivariate regression model of the PTEAM data. Thus, the estimates of the effect of smoking on indoor fine particle concentrations range from about 25–45 $\mu g/m^3$, with the higher value occurring in more northerly and therefore more insulated homes, with lower air exchange rates. Similar results were found in the largest U.S. study of particles in buildings (Turk et al. 1987; Turk et al. 1989); geometric means for PM3 were 44 and 15 $\mu g/m^3$ in smoking and nonsmoking areas, respectively.

d. Studies of ETS Exposure

The last and most recent of the four large-scale studies of particle exposures has been carried out over the years from about 1992 to the present (Heavner et al. 1995; Jenkins et al. 1994; Jenkins et al. 1995a; Jenkins et al. 1995b). About 100 nonsmoking persons in each of 16 U.S. cities, and between 190 and 225 persons in each of a planned ten European cities were recruited to carry a personal monitor for 24 hours, keep an activity diary, and supply saliva samples for cotinine determination. A front filter in the personal monitor collected particles, while a second acidified filter collected vapor-phase nicotine. The filters were analyzed for particle mass, nicotine, and various measures of combustion-related particles, including ultraviolet and fluorescence measurements as well as measurements of solanesol, a particle-bound organic substance that is specific to cigarettes. The particle size cut of the monitor was not well characterized, but appeared to extend as high as 50 μm diameter. This would correspond more nearly to total suspended particles (TSP) than to respirable particles (RSP). Thus, the mass of particles found ranged over an unusually high level, from a minimum of 20 to a maximum of 1,219 $\mu g/m^3$, with a mean value of 179 $\mu g/m^3$ and a median value of 142 $\mu g/m^3$. Persons with a smoking partner (N = 48) had mean exposures of 219 $\mu g/m^3$, compared to about 170 $\mu g/m^3$ for persons without a smoking partner (N = 207), a difference of about 49 $\mu g/m^3$. Using the solanesol marker for ETS, the corresponding values were 29 $\mu g/m^3$ ETS exposure for those with a smoking partner and 10 $\mu g/m^3$ for those without a smoking partner. Nicotine exposures averaged 4 $\mu g/m^3$ for those with a smoking partner and 1.3 $\mu g/m^3$ for those without. Thus, the mean values for both ETS and nicotine for persons with smoking spouses were only about three times higher than for those without smoking spouses. However, if the median values are employed, the ETS and nicotine exposures were about nine times higher for those with smoking spouses. Poor correlations of salivary cotinine with nicotine were reported (R^2 0.07–0.13). The authors provided five possible reasons, including the fact that saliva cotinine values were near the detection limit for many subjects. Another possible reason not mentioned by the authors was the possibility that the front filter might become acidified during collection of the sample, with the result that some of the nicotine could be trapped on that filter and thus escape detection. This was found to be the case in a study by Löfroth (1991) and also affected results from the earlier EPA PTEAM Study. On the other hand, quite a good correlation ($R^2 = 66\%$) was found between levels of nicotine and ETS particles.

2. VOCs

a. TEAM

Between 1979 and 1987, the EPA carried out the TEAM studies to measure personal exposures of the general public to VOCs in several geographic areas in the U.S. (Pellizzari 1987a; Pellizzari 1987b; Wallace 1987). About 20 target VOCs were included in the studies, which involved about 750 persons, representing 750,000 residents of the areas. Each participant carried a personal air quality monitor containing 1.5g Tenax. A small battery-powered pump pulled about 20L of air across the sorbent over a 12-hour period. Two consecutive 12-hour personal air samples were collected for each person. Concurrent outdoor air samples were also collected in the participants' back yards. In the studies of 1987, fixed indoor air samplers were also installed in the living rooms of the homes.

The initial TEAM pilot study (Wallace et al. 1982) in Beaumont, Texas, and Chapel Hill, North Carolina, indicated that personal exposures to about a dozen VOCs exceeded outdoor air levels, even though Beaumont has major oil producing, refining, and storage facilities. These findings were supported by a second pilot study in Bayonne-Elizabeth, New Jersey (another major chemical manufacturing and petroleum refining area) and Research Triangle Park, North Carolina (Wallace et al. 1984). A succeeding major study of 350 persons in Bayonne-Elizabeth (Wallace et al. 1984) and an additional 50 persons in a nonindustrial city and a rural area (Wallace et al. 1987a) reinforced these findings (see Table 8.1). A second major study in Los Angeles and in Antioch-Pittsburg, California (Wallace et al. 1988), with a follow-up study in Los Angeles in 1987 (Wallace et al. 1991), added a number of VOCs to the list of target chemicals with similar results (see Table 8.2).

Major findings of these TEAM Studies included the following:

1. Personal exposures exceeded median outdoor air concentrations by factors of two to five for nearly all of the 11 prevalent VOCs. The difference was even larger (factors of 10 or 20) when the maximum values were compared (see Figure 8.3). This is so despite the fact that most of the outdoor samples were collected in areas with heavy industry (New Jersey) or heavy traffic (Los Angeles).

Table 8.1 Weighted Arithmetic Mean Overnight Personal Exposures (Indoor Air) Compared to Outdoor Air Concentrations: New Jersey, All Three Seasons (μg/m³)

Chemical	Fall 1981 (128,000)ᵃ			Summer 1982 (109,000)			Winter 1983 (94,000)		
	Personal	Outdoor	I/O Ratio	Personal	Outdoor	I/O Ratio	Personal	Outdoor	I/O Ratio
1,1,1-Trichloroethane	110	5.4	20	21	10	2	31	1.4	22
m,p-Dichlorobenzene	56	1.5	37	49	1.4	35	54	1.2	45
m,p-Xylene	55	11	5	19	11	2	29	8.5	3
Tetrachloroethylene	11	3.7	3	9.0	4.0	2	13	1.9	7
Benzene	30	8.6	3	NCᵇ	NC	—	NC	NC	—
Ethylbenzene	13	3.8	3	7.8	3.5	2	11	3.4	3
o-Xylene	16	4.0	4	8.0	4.3	2	9.8	3.1	3
Trichloroethylene	7.3	2.1	3	4.8	7.8	0.6	3.0	0.2	15
Chloroform	8.7	1.2	7	4.6	12	0.4	4.0	0.1	15
Styrene	2.7	0.9	3	2.0	0.6	3	2.2	0.6	4
Carbon tetrachloride	14	1.2	12	1.2	1.0	1	NDᶜ	ND	—
Total (11 compounds)	324	43	7	126	56	2	157	20	8

ᵃ Population of Elizabeth and Bayonne, NJ, for which estimates apply.

ᵇ Not calculated — cartridges contaminated.

ᶜ Not detected (most samples).

Table 8.2 Weighted Estimates of Air and Breath Concentrations of 19 Prevalent Compounds

Chemical	Los Angeles (February 1984) (360,000)[a]			Los Angeles (May 1984) (330,000)			Contra Costa (June 1984) (91,000)		
	Personal Air (110)[b]	Outdoor Air (24)	Breath (110)	Personal Air (50)	Outdoor Air (23)	Breath (50)	Personal Air (76)	Outdoor Air (10)	Breath (67)
1,1,1-Trichloroethane	96[c]	34[c]	39[d]	44	5.9	23	16	2.8	16[d]
m,p-Xylene	28	24	3.5	24	9.4	2.8	11	2.2	2.5
m,p-Dichlorobenzene	18	2.2	5.0	12	0.8	2.9	5.5	0.3	3.7
Benzene	18	16	8.0	9.2	3.6	8.8	7.5	1.9	7.0
Tetrachloroethylene	16	10	12	15	2.0	9.1	5.6	0.6	8.6
o-Xylene	13	11	1.0	7.2	2.7	0.7	4.4	0.7	0.6
Ethylbenzene	11	9.7	1.5	7.4	3.0	1.1	3.7	0.9	1.2
Trichloroethylene	7.8	0.8	1.6	6.4	0.1	1.0	3.8	0.1	0.6
n-Octane	5.8	3.9	1.0	4.3	0.7	1.2	2.3	0.5	0.6
n-Decane	5.8	3.0	0.8	3.5	0.7	0.5	2.0	3.8	1.3
n-Undecane	5.2	2.2	0.6	4.2	1.0	0.7	2.7	0.4	1.2
n-Dodecane	2.5	0.7	0.2	2.1	0.7	0.4	2.1	0.2	0.4
α-Pinene	4.1	0.8	1.5	6.5	0.5	1.7	2.1	0.1	1.3
Styrene	3.6	3.8	0.9	1.8	—	—	1.0	0.4	0.7
Chloroform	1.9	0.7	0.6	1.1	0.3	0.8	0.6	0.3	0.4
Carbon tetrachloride	1.0	0.6	0.2	0.8	0.7	0.2	1.3	0.4	0.2
1,2-Dichloroethane	0.5	0.2	0.1	0.1	0.06	0.05	0.1	0.05	0.04
p-Dioxane	0.5	0.4	0.2	1.8	0.2	0.05	0.2	0.1	0.2
o-Dichlorobenzene	0.4	0.2	0.1	0.3	0.1	0.04	0.6	0.07	0.08
TOTAL (19 Compounds)	240	120	80	150	33	56	72	16	44

a Population for which estimates apply.
b Number of 12-hour samples.
c Average of arithmetic means of day and night 12-hour samples (µg/m³).
d One very high value removed.

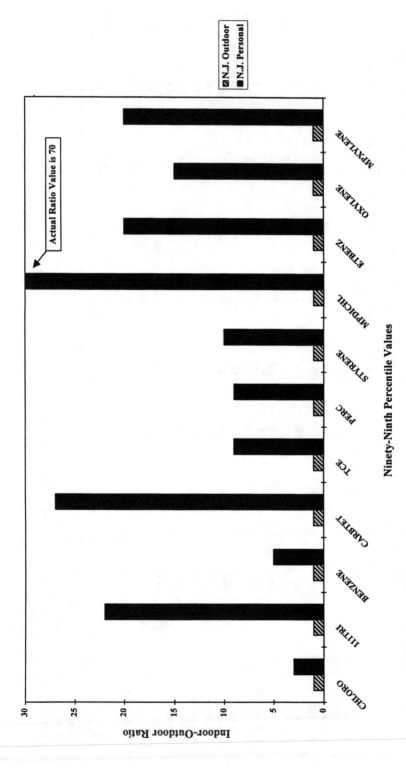

Figure 8.3   Comparison of unweighted 99th percentile concentrations of 11 prevalent chemicals in overnight outdoor air and overnight personal air in New Jersey (Fall 1981).

2. Major sources are consumer products (bathroom deodorizers, moth repellents); personal activities (smoking, driving); and building materials (paints and adhesives). In the U.S., one chemical (carbon tetrachloride) has been banned from consumer products and exposure is thus limited to the global background of about 0.7 $\mu g/m^3$.

3. Traditional sources (automobiles, industry, petrochemical plants) contributed only 20–25% of total exposure to most of the target VOCs (Wallace 1991a, 1991b). No difference in exposure was noted for persons living close to chemical manufacturing plants or petroleum refineries.

A more recent study of benzene and toluene in 293 California homes (Wilson et al. 1993a; Wilson et al. 1993b; Wilson et al. 1995a; Colome et al., 1994) resulted in some interesting differences between the two agents. For benzene, 48-hour average indoor concentrations correlated fairly well with outdoor levels, but almost no correlation was observed for toluene. This is likely due to the much wider use of toluene in consumer products. Major variables associated with higher net indoor benzene levels were presence of a gas furnace and having two cars parked in an attached garage. For toluene, a particular brand of furnace had the highest partial correlation with net indoor toluene concentrations; apartments also had higher concentrations.

Another study of benzene in four New Jersey homes focused on the extent of contamination from attached garages (Thomas et al. 1993b). Each home was monitored for either six or ten consecutive 12-hour periods. At all four homes, garage levels of benzene were higher than outdoors, and at three of the four homes the garage levels were higher than in the living area. Air exchange measurements made it possible to calculate the amount of benzene transferred from the garage to the living area in the four homes; in the home without elevated benzene levels in the garage, the total air flow between the garage and the living area was extremely small. Benzene concentrations in the garages ranged from 5–200 $\mu g/m^3$, and the 12-hour average source strength estimates ranged from 730 to 26,000 $\mu g/h$. The mere presence of an attached garage was not a significant factor in affecting living area benzene concentrations; however, the total number of hours the car was parked in the garage had a significant effect on living area benzene concentrations, as did the mass flow rate of benzene from the garage to the home.

During the most recent National Health and Nutrition Examination Study (NHANES), the isotope dilution method described above was employed to look for about 20 VOCs in approximately 1,000 persons representing the U.S. population (Ashley et al. 1994). Essentially the same dozen or so VOCs observed in the TEAM studies were prevalent in the blood of the respondents.

A study of personal exposure to benzene was carried out in Valdez, Alaska, to determine the contribution of the major petroleum terminal (Goldstein et al. 1992). Fifty-eight residents wore Tenax personal samplers for one day in summer and winter, while simultaneous indoor and outdoor measurements were made in or near their homes. To estimate the impact of the terminal, a tracer gas (SF_6) was emitted continuously from the terminal. The amount of SF_6 on the residents' charcoal collectors could be used to estimate the proportion of benzene exposure contributed by the terminal. This proportion was found to be 11%, even though the terminal accounted for at least 90% of the total atmospheric emissions of benzene.

Three large studies of VOCs, involving 300 to 800 homes, were carried out in the Netherlands (Lebret et al. 1986), West Germany (Krause et al. 1987; Seifert et al. 1987), and the U.S. (Wallace 1987). A smaller study of 15 homes was carried out in Northern Italy (De Bortoli et al. 1986). Observed concentrations were remarkably similar for most chemicals, indicating similar sources in these countries. One exception is chloroform, which was present at typical levels of 1–4 $\mu g/m^3$ in the U.S. but not found in European homes. This is to be expected, since the likely source is volatilization from chlorinated water (Wallace et al. 1982; Andelman, 1985a; Andelman 1985b; Andelman 1990) and the two European countries do not chlorinate their water.

Major findings of these indoor air studies include:

1. Indoor levels in homes and older buildings (greater than one year in age) are typically several times higher than outdoor levels. Sources include dry-cleaned clothes (Thomas et al. 1991), cosmetics, air fresheners, and cleaning materials.
2. New buildings (less than one month in age) have levels of some VOC (aliphatics and aromatics) 100 times higher than outdoor, falling to ten times outdoor about two to three months later. Major sources include paints and adhesives.
3. About half of 750 homes in the U.S. had total VOC levels (obtained by integrating the total ion current response curve of the mass spectrometer) greater than 1 mg/m^3, compared to only 10% of outdoor samples (Wallace et al. 1991a).
4. More than 500 volatile organic compounds (VOCs) were identified in four buildings in Washington, D.C. and Research Triangle Park, North Carolina (EPA 1988a).

One study (Wallace et al. 1989) involved seven volunteers undertaking about 25 activities suspected of causing increased VOC exposures; a number of these activities (using bathroom deodorizers, washing dishes, cleaning an auto carburetor) resulted in 10- to 1,000-fold increases in eight-hour exposures to specific VOCs.

A recent study of 170 homes in Avon, England found mean indoor levels of benzene to be 8 $\mu g/m^3$, compared to outdoor concentrations of 5 $\mu g/m^3$ (Brown and Crump, 1996). The study employed passive Tenax tubes to collect 28-day indoor and outdoor samples. These results were in agreement with the levels of 10 $\mu g/m^3$ indoors and 6 $\mu g/m^3$ outdoors at 50 homes in Los Angeles measured over two seasons in 1987 (Wallace et al. 1991a).

Kostiainen (1995) quantified 48 VOCs in 38 complaint homes and 50 control homes, finding that levels were generally higher in the complaint homes. Aromatics, terpenes, alkylcyclohexanes, and two halocarbons had the highest frequency of elevated concentrations in the sick homes. Case studies indicated the sources included gasoline spills, renovation of a tetrachloroethylene-using laundry into a residence, a ground floor parking garage, and a leather sofa that emitted 1,1,1-trichloroethane, trichloroethylene, p-dichlorobenzene, and 1-acetoxy-2-ethoxy-ethane in high concentrations.

b. Soil-Gas Transport

A transport mode of particular interest in assessing residential exposures is the movement of soil gas into basements. This is of particular interest in situations involving plumes of hydrocarbons emanating from gasoline spills, hazardous waste sites, and leaking underground storage tanks. Fundamental physical theory and experiments have been developed, notably by the Indoor Air Program at Lawrence Berkeley Laboratories, to determine the important variables affecting soil-gas transport, as well as corrective actions that may be taken (Hodgson et al. 1988). It appears that the same forces (e.g., negative pressure) that allow radon to enter buildings in soil gas can also result in VOCs and perhaps some termiticides entering. Some of the same protective actions against radon (e.g., rerouting soil gas into the atmosphere) may also reduce entry of VOCs.

A building on a site contaminated by a gasoline spill was studied by Fischer et al. (1996). Soil gas measurements 0.7 m below the surface had high VOC concentrations of 30–60 g/m³. These concentrations were reduced by about a factor of 1,000 as the gas moved to the surface and by another factor of 1,000 as it was diluted by surface air ventilation. However, the authors caution that it is not known whether these results can be applied to other sites.

c. Sources

Early studies of organics indoors were carried out in the 1970s in the Scandinavian countries (Johansson 1978; Mølhave and Møller 1979; Berglund et al. 1982a; Berglund et al. 1982b). Mølhave (1982) showed that many common building materials used in Scandinavian buildings emitted organic gases. Seifert and Abraham (1982) found higher levels of benzene and toluene associated with storage of magazines and newspapers in German homes. Early U.S. measurements were made in nine Love Canal residences (EPA 1979); 34 Chicago homes (Jarke and Gordon 1981); and in several buildings (Hollowell and Miksch 1981; Miksch, Hollowell, and Schmidt 1982).

Hundreds of VOCs have been identified in environmental tobacco smoke (Higgins 1987; Guerin et al. 1987; Jermini et al. 1976; Löfroth et al. 1989) found in about 60% of all U.S. homes and workplaces (Repace and Lowrey 1980; Repace and Lowrey 1985) in the mid-1980s, but probably closer to 40% today. Among these are several human carcinogens, including benzene. Benzene was elevated in the breath of smokers by a factor of 10 above that in the breath of nonsmokers (Wallace et al. 1987b). The amount of benzene in mainstream smoke appears to be directly related to the amount of tar and nicotine in the cigarette (Higgins et al. 1983). In the U.S., it is calculated that the 50 million smokers are exposed to about half of the total nationwide "exposure budget" for benzene (Wallace 1989; Wallace 1990). Other indoor combustion sources such as kerosene heaters (Traynor et al. 1990) and woodstoves (Highsmith et al. 1988) may emit both volatile and semivolatile organic compounds.

Later studies also investigated building materials (EPA 1988a, 1988b) but added cleaning materials and activities such as scrubbing with chlorine bleach or spraying insecticides (Wallace et al. 1987c) and using adhesives (Girman et al. 1987) or paint removers (Girman and Hodgson 1987). Knöppel and Schauenburg (1987) studied VOC emissions from ten household products (waxes, polishes, detergents); 19 alkanes, alkenes, alcohols, esters, and terpenes were among the chemicals emitted at the highest rates from the ten products. All of these studies employed either head-space analysis or chambers to measure emission rates.

Other studies estimated emission rates from measurements in homes or build-ings. For example, Wallace (1987) estimated emissions from a number of personal activities (e.g., visiting dry cleaners and pumping gas) by regressing measurements of exposure or breath levels against the specified activities. Girman and Hodgson (1987) extended their chamber studies of paint removers to a residence, finding similar (and very high ppm) concentrations of methylene chloride in this more realistic situation.

The U.S. National Aeronautics and Space Agency (NASA) measured organic emissions from about 5,000 materials used in space missions (Nuchia 1986). Perhaps 3,000 of these materials are in use in general commerce (Özkaynak et al. 1987). The chemicals emitted from the largest number of materials included toluene (1,896 materials), methyl ethyl ketone (1,261), and xylenes (1,111).

A 41-day chamber study (Berglund et al. 1987) of aged building materials taken from a "sick" preschool indicated that the materials had absorbed about thirty VOCs, which they reemitted to the chamber during the first 30 days of the study. Only 13 of the VOCs originally present in the first days of the study continued to be emitted in the final days, indicating that these 13 were the only true components of the materials. This finding has significant implications for remediating "sick buildings." Even if the source material is identified and removed, weeks may be needed before reemission of organics stops from materials in the building.

Emission rates of most chemicals in most materials are greatest when the mate-rials are new. For "wet" materials such as paints and adhesives, most of the total volatile mass may be emitted in the first few hours or days following application (Tichenor and Mason 1987; Tichenor et al. 1990). The EPA studies of new buildings indicated that eight of 32 target chemicals measured within days after completion of the building were elevated 100-fold compared to outdoor levels: xylenes, ethyl-benzene, ethyltoluene, trimethylbenzenes, decane, and undecane (EPA 1988b). The half-lives of these chemicals varied from 2 to 6 weeks; presumably some other nontarget chemicals, such as toluene, would have shown similar behavior. The main sources were likely to be paints and adhesives. Thus, new buildings would be expected to require about 6 months to a year to decline to the VOC levels of older buildings.

For dry building materials, such as carpets and pressed wood products, emissions are likely to continue at low levels for longer periods. Formaldehyde from pressed wood products may be slowly emitted with a half-life of several years (Breysse 1984). According to several recent studies, 4-phenylcyclohexene (4-PC), a reaction product occurring in the styrene-butadiene backing of carpets, is the main VOC emitted from carpets after the first few days. This material is likely to be largely responsible for the new carpet odor.

A major category of human exposure to toxic and carcinogenic VOCs is room air fresheners and bathroom deodorants. Since the function of these products is to maintain an elevated indoor air concentration in the home or the office over periods of weeks (years with regular replacement), extended exposures to the associated

VOCs are often the highest likely to be encountered by most (nonsmoking) persons. The main VOCs used in these products are paradichlorobenzene (widely used in public restrooms), limonene (lemon scent), and α-pinene (pine scent). The first is carcinogenic to two species (NTP 1986), the second to one (NTP, 1988), and the third is mutagenic. Limonene and α-pinene also are used in many cleaning and polishing products, which would cause short-term peak exposures during use but which might not provide as much total exposure as the air freshener.

Awareness is growing that most exposure comes from these small nearby sources. In California, Proposition 65 focuses on consumer products, requiring makers to list carcinogenic ingredients. Environmental tobacco smoke was declared a known human carcinogen by the EPA in 1991; smoking has been banned from many public places and many private workplaces during the last few years.

d. Pesticides

The main study of personal exposure and indoor concentrations of pesticides was sponsored by the EPA in 1984–86 (EPA 1989b, 1990c, 1990d). Two cities, one with expected high pesticide use (Jacksonville, Florida) and one with expected low use (Springfield, Massachusetts) were chosen. About 250 persons, selected to represent the populations of the two cities, were monitored for one day (24 hours). Personal, indoor, and concurrent outdoor concentrations were measured for about 32 target pesticides using a sampler containing polyurethane foam (PUF) as a sorbent.

Major findings included the following:

1. Both personal exposures and indoor concentrations were much greater than outdoor concentrations for all target pesticides. In most cases, indoor sources contributed well over 90% of total exposure.
2. The pesticides with the highest cancer risks were the chlorinated hydrocarbons such as chlordane, heptachlor, aldrin, and dieldrin. Although all these pesticides had been banned for some years, they were still present in indoor air, presumably due to volatilization from termite-treated soil.

e. PCBs

The presence of PCBs in indoor air is probably more common than generally believed, partly because the methodology for measurement is expensive and not many measurements have been made. The 1977 ban of PCBs did not extend to existing closed-system electrical devices such as transformers and ballasts for fluorescent lights. An interesting observation was made by Wallace, Basu, and Hites (1996), who discovered that earlier measurements of outdoor PCBs were contaminated by indoor PCBs, either from rooftop ventilation during sampling or in the laboratories where samples were prepared for analysis. The authors recommended measuring indoor air for PCBs in buildings where rooftop samplers are located and in laboratories where samples are prepared for analysis.

3. Inorganic Gases

a. Carbon Monoxide

The largest personal monitoring study of CO exposures was carried out by the EPA in Washington, D.C., and Denver, Colorado, in the winter of 1982–83 (Akland et al. 1985; EPA 1984; EPA 1983, 1986). About 800 persons in D.C. and 450 in Denver were monitored for 24 hours (48 hours in Denver) using electrochemical CO monitors with specially designed data loggers. The data loggers were capable of sampling the current from the CO monitor about four times a second. They were equipped with buttons that the subject could press when one activity ended and the next began; at that point, the logger would average all preceding values from the time the activity began; there was also an automatic averaging every hour on the hour. The result was an extraordinarily rich data base, with approximately 1,200 persons averaging 40 activities per day, each with an associated CO average. At the end of the monitoring period, each subject provided a breath sample.

Major findings of the study included the following:

1. Commuters had the highest exposures to CO in general, averaging up to 13 ppm. Parking garages had the highest CO levels of any microenvironment, with churches and schools among the lowest.
2. The main indoor sources of CO were gas stoves and cigarettes. Gas stoves increased levels by about 2.5 ppm when being used; homes with smokers had increases of about 1.5 ppm on average.
3. Personal exposures were higher than would be predicted by the fixed stations. About 10% of D.C. residents appeared to exceed the 8-hour outdoor CO standard of 9 ppm, as determined by their breath concentrations, although only one of the 11 fixed stations exceeded the CO standard during the monitoring period.

A study of California homes (Wilson et al. 1993a; Wilson et al. 1993b; Wilson et al. 1995a; Colome et al. 1994), each monitored for 48 hours, indicated that 13 of 277 homes (about 5%) had indoor 8-hour averages exceeding the 9 ppm CO outdoor standard. Since the outdoor standard may be exceeded only once per year, it is clear that the fraction of homes with 8-hour indoor averages exceeding 9 ppm more than once per year would be larger than the 5% observed in the single 48-hour monitoring period. Homes with gas stoves and gas furnaces had indoor source terms for CO that were about three times higher than homes without such sources. Homes with wall furnaces had higher CO levels than homes with forced-air gas furnaces. Homes with smokers (N = 85) had CO levels about 0.5 ppm higher than homes without smokers (N = 190). Malfunctioning gas furnaces were a major cause of elevated CO concentrations. However, the homes with the highest CO levels also included some with electric stoves and electric heat, suggesting that other sources of CO were present in these homes. Such sources could include cars idling in attached garages or unvented gas or kerosene space heaters.

b. Nitrogen Dioxide

NO_2, like CO and particles, is an EPA criteria air pollutant with a NAAQS. NO_2 is emitted by industrial processes, but also by indoor combustion appliances such as gas stoves and furnaces. Several studies in the 1970s (Goldstein et al. 1979) suggested that children in homes with gas stoves suffered more infectious disease than children in homes with electric stoves; a possible connection with NO_2 (in lowering resistance) was postulated. Therefore, several large-scale population-based studies of NO_2 in homes have been carried out in the past two decades.

An early study was sponsored by the Southern California Gas Corporation (Colome et al. 1987; Wilson et al. 1986). More than 500 homes served by the company were selected to participate in a study of week-long NO_2 concentrations in homes. The homes were visited during three seasons: spring and summer of 1984 and winter (January) of 1985. Major findings included:

1. Homes with gas stoves had higher concentrations than homes with electric stoves.
2. Homes with gas-fired wall furnaces had higher concentrations than homes with gas floor furnaces.
3. Pilot lights accounted for about one-third of total NO_2 emissions from gas stoves so equipped.
4. In homes with gas stoves, kitchens had higher concentrations than living rooms, which in turn had higher concentrations than bedrooms.

Additional major studies of indoor concentrations of, and personal exposures to, NO_2 took place in Boston (Ryan et al. 1988a; Ryan et al. 1988b; Ryan et al. 1992) and Los Angeles (Spengler et al. 1992; Schwab et al. 1990). Personal exposures were found to be well correlated with bedroom measurements ($R^2 = 0.48$ in Boston, 0.53 in Los Angeles), and also reasonably well correlated with outdoor concentrations in Los Angeles ($R^2 = 0.41$). However, in Boston, which has lower outdoor NO_2 than Los Angeles, personal exposures were poorly correlated with outdoor concentrations ($R^2 = 0.09$). The best model for personal exposure included season and cooking fuel in Boston, and season and range/furnace type in Los Angeles. In both cities, persons with gas ranges had about 10 ppb higher exposures than those without gas ranges.

B. Air Exchange

The largest study to measure air exchange rates in homes was the Southern California Gas Corporation study of NO_2 in southern California homes (Wilson et al. 1986). In this study, NO_2 was monitored indoors and outdoors for one week at 597 homes in March of 1984, 444 of those homes in July of 1984, and 405 of the same homes in January of 1985.

Another large study was the New York State study of particles mentioned above; more than 400 homes in Suffolk and Onondaga County were monitored for particle levels over a period of one week; week-long average air exchange rates were measured simultaneously.

The EPA PTEAM Study, discussed above, measured daytime and overnight air exchange rates in 178 homes in Riverside, California. The EPA 1987 TEAM Studies in Baltimore, Maryland, and Los Angeles, California, also measured air exchange rates in 150 Baltimore homes and in 50 Los Angeles homes, the latter over two seasons.

VI. MODELS

Mass balance models have been used for more than a century in various branches of science. All such models depend on the law of the conservation of mass. They simply state that the change in mass of a substance in a given volume is equal to the amount of mass entering that volume minus the amount leaving the volume. Usually they are written in the form of first-order linear differential equations. For example, consider a volume V filled with a gas of mass m. The change in mass Δm over a small time Δt will be the difference between the mass entering the volume (m_{gain}) and the mass leaving the volume (m_{loss}):

$$\Delta m/\Delta t = (m_{gain} - m_{loss})/\Delta t \tag{1}$$

Taking the limit as Δt approaches zero, we have the differential equation for the rate of change of the mass:

$$dm/dt = d/dt\,(m_{gain} - m_{loss}) \tag{2}$$

If we require that the mass be uniformly distributed throughout the volume at all times, we have a condition that physical chemists call "well-mixed." We assume that any mass gained or lost in the volume V is instantaneously distributed evenly throughout the volume. We may then replace the mass terms by the concentration $C = m/V$:

$$VdC/dt = d/dt\,(m_{gain} - m_{loss}) \tag{3}$$

The above equation is the basis for all mass-balance models dealing with well-mixed compartments. It takes on many forms depending on the type of processes involved in transporting mass into or out of the volume being considered. A large class of models assume that the volume is a single compartment. More complex models assume multiple compartments. As an example of a single compartment model, we may consider a room of volume V that exchanges air with the outside at a constant flow rate Q. We also assume that a mass of gas has been released in the room at time $t = 0$, and that the outdoor concentration of this gas is 0. This is the situation, for example, when a tracer gas such as SF_6 is released to determine the air exchange rate. In this case, the gain in mass (m_{gain}) is 0 and the loss in mass is equal to the flow rate Q out of the house times the concentration C, so that equation (3) becomes:

$$VdC/dt = -QC \qquad (4)$$

Integrating this equation by separation of variables, we have:

$$C = C_0 e^{-at} \qquad (5)$$

where $a = Q/V$ is the air exchange rate, and C_0 is the concentration at time $t = 0$. Thus, we find that the original concentration of the tracer in the room decays with a time constant a: the air exchange rate.

For a nonreactive gas with a nonzero outdoor concentration (e.g., carbon monoxide), the mass balance equation takes the form:

$$dC_{in}/dt = a(C_{out} - C_{in}) \qquad (6)$$

Depending on the variation with time of C_{out}, this equation has a number of solutions. If C_{out} is constant, for example, and the initial indoor concentration is zero, then the indoor concentration rises at a rate determined by the air exchange rate to approach an asymptotic value equal to the outdoor concentration:

$$C_{in} = C_{out}(1 - e^{-at}) \qquad (7)$$

An early effort at developing an indoor air quality model was made by Shair and Heitner (1974). This was a mass balance model in which the building was represented as a single well-mixed chamber. A single first-order linear differential equation represented the change in mass of an agent due to infiltration, exfiltration, recirculation, source generation, and removal due to filters in the circulation system or deposition on surfaces. Shair and Heitner solved the equation for certain simple inputs, such as a linearly increasing or decreasing outdoor concentration:

$$C_{out} = mt + b \qquad (8)$$

Since the outdoor concentration is normally a slowly varying function, Shair and Heitner's linear approximation is actually an excellent approximation for time intervals of moderate length.

If an indoor source $S(t)$ exists, it enters the mass balance model in the following way:

$$dC_{in}/dt = a(C_{out} - C_{in}) + S(t)/V \qquad (9)$$

where $S(t)$ has the units of mass per unit time.

If the source has a constant generation rate (e.g., CO_2 emissions from a person at rest), then $S(t)$ is a constant value S_0 and the equation becomes:

$$dC_{in}/dt = a(C_{out} - C_{in}) + S_0/V \qquad (10)$$

If the substance of interest reacts or is adsorbed on surfaces while indoors, the equation becomes

$$dC_{in}/dt = aC_{out} - (a + k)C_{in} + S(t)/V \qquad (11)$$

where k represents the loss of the substance due to chemical reaction, adsorption on surfaces, sedimentation, etc. The decay rate k has the same units as the air exchange rate a (1/time); their sum $(a + k)$ may be thought of as an effective air exchange rate. The decay rate k is often used to apply to particles, which disappear more quickly indoors than a nonreactive gas such as CO. Since particles may experience more difficulty than molecules of a gas in penetrating the building envelope, a penetration factor P is often applied that multiplies the outdoor concentration as in equation (11) above.

As described above, Koutrakis et al. (1992) used least-squares analysis to solve a simplified form of the mass-balance model to determine source emission rates for particles and elements due to cigarettes, woodsmoke, and kerosene heater use. Koutrakis assumed a value for k in order to solve the equation for P and the source emission rates. Özkaynak et al. (1993) improved on Koutrakis's approach by using nonlinear least-squares analysis of the PTEAM results to solve the equation simultaneously for k, P, and source emission rates for $PM_{2.5}$ and PM_{10} particles and associated elements for smoking and for cooking.

Traynor et al. (1989) developed a "macromodel" based on Monte Carlo simulations using global input data such as house volumes, air exchange rates, and emissions from combustion sources to assess residential concentrations of combustion-source agents such as CO, NO_2, and respirable suspended particles. For a home with only one combustion source during winter in upstate New York, at an outdoor temperature of 0°C, and an outdoor RSP geometric mean concentration of 19 μg/m³, the model predicted geometric mean concentrations of about 80 μg/m³ in a home with smoking, 75 μg/m³ for a radiant kerosene heater, about 60 μg/m³ for a convective unvented gas space heater and a nonairtight wood stove, and about 40 μg/m³ for an infrared unvented gas space heater. An airtight wood stove was predicted to produce a geometric mean about 30 μg/m³. Gas ovens, dryers, hot water heaters, boilers, and forced-air furnaces were predicted to result in low indoor concentrations of 10–15 μg/m³, unless the gas oven was used for heating, in which case the predicted geometric mean was about 20 μg/m³.

At present, the most complete form of the mass-balance indoor air quality model was presented by Nazaroff and Cass (1989). These authors developed the model to allow for changes in particle size and chemical composition, including terms for homogenous turbulence, natural convection, thermophoresis, advection, and Brownian motion. Coagulation of particles is also included. The computer form of the model required 40 to 60 minutes of CPU time to determine an 11-hour evolution of an aerosol mixture of 16 different sizes. The model was partially validated by checking it against the results of a chamber study using cigarette-generated aerosol to determine the effectiveness of air cleaners (Offermann et al. 1985).

A crucial unknown parameter in the mass-balance model for particles is the rate of decay to surfaces. This rate of decay (k) may be expressed as the product of a deposition velocity k_d with the surface-to-volume ratio in the room or building:

$$k = k_d S/V \tag{12}$$

The deposition velocity will vary with particle size.

Both the Nazaroff study above and the series of studies by Weschler and colleagues below have provided useful data on deposition velocities for important anions such as sulfates.

A series of studies, also concerned with the effects of indoor particles on materials, were carried out by Weschler and colleagues at AT&T Bell Laboratories (Weschler et al. 1989; Sinclair et al. 1988; Sinclair et al. 1990; Sinclair et al. 1992). Studies of buildings with low occupancy, large amounts of electronic equipment, and high-quality filtering and HVAC systems succeeded in determining deposition velocities for coarse particles and various fine particle ions. For coarse particles, these velocities were about equal to velocities predicted for gravitational settling, similar to the results of Nazaroff et al. (1990a) described above. For fine particles, however, the deposition velocity was greater than that predicted for gravitational settling alone. For sulfates, the dominant anion in fine particles, deposition velocities at four buildings in Wichita, Lubbock, Newark, and Neenah were 0.004, 0.005, 0.005, and 0.004 cm/s, respectively (Sinclair et al. 1992).

Nazaroff et al. (1993) reviewed these and other studies of deposition velocity. The authors pointed out that the studies by Weschler and colleagues and also one study in Helsinki (Raunemaa et al. 1989) had produced values of 0.003–0.005 cm/s for fine-mode sulfate, but that studies by Nazaroff and colleagues (Nazaroff et al. 1990a) resulted in much smaller values of 0.00002–0.001 cm/s. It is not clear whether the differences are due to the many differences in surface materials and filtration systems in the different types of buildings (museums vs. telephone equipment buildings) or to the different methods of determining deposition velocities. However, the discrepancy is clear evidence that further work is needed.

Indoor air models for buildings have undergone considerable international development and testing. An important recent model is COMIS, developed by scientists at Lawrence Berkeley National Laboratory. The model requires quite a bit of input data, including wind speed and direction, pressure, temperature, leakage characteristics, dimensions of doors and windows, fan characteristics, layout of duct work, pollutant source strengths within the buildings, and pollutant concentrations in ambient air. The model has been extensively evaluated, with more than 50 benchmarks tested against existing analytical or numerical solutions. An international user test was also performed, which helped test and improve the documentation. Finally nine *in situ* tests were performed and compared with the model predictions.

An indoor air quality model called CONTAM has been developed at the National Institute for Standards and Technology. CONTAM96, the latest version, is obtainable via the Internet. Input requirements for CONTAM are much the same as in COMIS. Air flows and contaminant concentrations are calculated for each zone, and all equations are solved simultaneously for all zones and contaminants.

A model called EXPOSURE, providing estimates of personal exposure within a residence, was developed at the EPA (Sparks et al. 1991; Sparks et al. 1993). This model is considerably less complicated than COMIS or CONTAM, and places somewhat more emphasis on pollutant sources and personal activity patterns.

Bogen and McKone (1988) linked an indoor air model to a pharmacokinetic model to assess risk from tetrachloroethylene. They found that time-weighted averages of indoor air concentrations together with a steady-state PBPK model yielded estimates of total metabolized tetrachloroethylene similar to those obtained using the far more complicated continuous dynamic modeling.

BIBLIOGRAPHY

Akland, G., Hartwell, T.D., et al. 1985. Measuring human exposure to carbon monoxide in Washington, DC, and Denver, CO, during the winter of 1982–83, *Environmental Science and Technology* 19:911–918.

Andelman, J.B. 1985a. Human exposures to volatile halogenated organic chemicals in indoor and outdoor air, *Environmental Health Perspectives* 62:313–318.

Andelman, J.B. 1985b. Inhalation exposure in the home to volatile organic contaminants of drinking water, *Sci. Total Environ.* 47:443–460.

Andelman, J.B. 1990. Total exposure to volatile organic compounds in potable water, In: *Significance and Treatment of Volatile Organic Compounds in Water Supplies,* Ram, N.M., Christman, R.F., Cantor, K.P., Eds., Lewis Publishers: Chelsea, MI.

Ashley, D.L., Bonin, M.A., et al. 1994. Blood concentrations of volatile organic compounds in a nonoccupationally exposed U.S. population and in groups with suspected exposure, *Clinical Chemistry* 40:1401–14.

Barkley, J., Bunch, J., et al. 1980. Gas chromatography mass spectrometry computer analysis of volatile halogenated hydrocarbons in man and his environment—a multimedia environmental study, *Biomedical Mass Spectrometry* 7(4):139–147.

Berglund, B., Johansson, I., et al. 1982a. A longitudinal study of air contaminants in a newly built preschool, *Environment International* 8:111–115.

Berglund, B., Johansson, I., et al. 1982b. The influence of ventilation on indoor/outdoor air contaminants in an office building, *Environment International* 8:395–399.

Berglund, B., Johansson, I., et al. 1987. Volatile organic compounds from building materials in a simulated chamber study, In: *Proceedings of the 4th International Conference on Indoor Air Quality and Climate, Vol. 1,* Institute for Soil, Water and Air Hygiene, Berlin: 16.

Bogen, K.T., McKone, T.E. 1988. Linking indoor air and pharmacokinetic models to assess tetrachloroethylene risk, *Risk Analysis* 8(4):509–520.

Breysse, P. A. 1984. Formaldehyde levels and accompanying symptoms associated with individuals residing in over 1000 conventional and mobile homes in the state of Washington, In: *Indoor Air: Sensory and Hyperreactivity Reactions to Sick Buildings, Volume 3,* Berglund, B. Lindvall, T., Sundell, J., Eds., Swedish Council for Building Research: Stockholm, Sweden, 403–408

Brown, V.M., Crump, D.R. 1996. Volatile organic compounds, In: *Indoor Air Quality in Homes: Part I. The Building Research Establishment Indoor Environment Study,* Berry, R.W., Brown, V.M., Coward, S.K.D., et al., Eds., Construction Research Communications: London, England.

Clayton, C.A., Perritt, R.L., et al. 1993. Particle total exposure assessment methodology (PTEAM) study: Distributions of aerosol and elemental concentrations in personal, indoor, and outdoor air samples in a Southern California community, *Journal of Exposure Analysis and Environmental Epidemiology* 3:227–250.

Colome, S.D., Wilson, A.L., et al. 1987. Analysis of factors associated with indoor residential nitrogen dioxide: Multivariate regression results, In: *Indoor Air 87, Vol. I, Proceedings of the 4th International Conference on Indoor Air,* Institute for Water, Soil and Air Hygiene, Seifert, B. et al., Eds. Berlin: 405–409.

Colome, S.D., Wilson, A.L., et al. 1994. *California Residential Indoor Air Quality Study, Volume 2: Carbon Monoxide and Air Exchange Rate: A Univariate & Multivariate Analysis,* Integrated Environmental Services: Irvine, CA.

Coutant, R.W., Lewis, R.G., et al. 1985. Passive sampling devices with reversible adsorption, *Analytical Chemistry* 57:219–223.

Coutant, R.W., Lewis, R.G., et al. 1986. Modification and evaluation of a thermally desorbable passive sampler for volatile organic compounds in air, *Analytical Chemistry* 58:445–448.

De Bortoli, M., et al. 1986. Concentrations of selected organic pollutants in indoor and outdoor air in northern Italy, *Environment International* 12(1–4):343–350.

De Koster, J.A., Thorne, P.S. 1995. Bioaerosol concentrations in noncompliant, compliant, and intervention homes in the midwest, *American Industrial Hygiene Association Journal* 56:573–580.

Dietz, R.N., Cote, E.A. 1982. Air infiltration measurements in a home using a convenient perfluorocarbon tracer technique, *Environment International* 8:419–33.

Dockery, D.W., Schwartz, J., et al. 1992. Air pollution and daily mortality: Associations with particulates and acid aerosols, *Environmental Research* 59:362–373.

Dockery, D.W., Spengler, J.D. 1981a. Indoor-outdoor relationships of respirable sulfates and particulates, *Atmospheric Environment* 15:335–343.

Dockery, D.W., Spengler, J.D. 1981b. Personal exposure to respirable particulates and sulfates, *Journal of the Air Pollution Control Association* 31(2):153–159.

Duan, N. 1991. Stochastic microenvironment models for air pollution exposure, *Journal of Exposure Analysis and Environmental Epidemiology* 2:235–257.

Environmental Protection Agency (EPA). 1979a. *Analysis of Organic Air Pollutants by Gas Chromatography and Mass Spectroscopy,* U.S. Environmental Protection Agency, Research Triangle Park, NC.

Environmental Protection Agency (EPA). 1979b. *Formulation of a Preliminary Assessment of Halogenated Organic Compounds in Man and Environmental Media,* U.S. Environmental Protection Agency, Washington, DC.

Environmental Protection Agency (EPA). 1983. *A Study of Personal Exposures to Carbon Monoxide in Denver, CO,* Final Report, U.S. Environmental Protection Agency, Research Triangle Park, NC. Contract # 68-02-3755.

Environmental Protection Agency (EPA). 1984. *Study of Carbon Monoxide Exposure of Residents of Washington, DC, and Denver, CO,* EPA-600/S4-84-031, U.S. Environmental Protection Agency, Environmental Monitoring Systems Laboratory, Research Triangle Park, NC.

Environmental Protection Agency (EPA). 1986. *Selected Data Analyses Relating to Studies of Personal Carbon Monoxide Exposure in Denver and Washington, DC,* Final Report, U.S. Environmental Protection Agency, Research Triangle Park, NC. Contract #68-02-3496.

Environmental Protection Agency (EPA). 1988a. *Indoor Air Quality in Public Buildings: Vol. I*, EPA 600/6-88/009a, U.S. Environmental Protection Agency, Washington, DC.

Environmental Protection Agency (EPA). 1988b. *Indoor Air Quality in Public Buildings: Vol. II*, EPA 600/6-88/009b, U.S. Environmental Protection Agency, Research Triangle Park, NC.

Environmental Protection Agency (EPA). 1989a. *Indoor Air Quality and Work Environment Survey: EPA Headquarters Buildings, Volume I: Employee Survey*, EPA Contract # 68-01-7359, U.S. Environmental Protection Agency, Washington, DC.

Environmental Protection Agency (EPA). 1989b. *Nonoccupational Pesticide Exposure Study (NOPES)*, Final Summary Report, U.S. Environmental Protection Agency, Research Triangle Park, NC. EPA Contract # 68-02-4544.

Environmental Protection Agency (EPA). 1990a. *Indoor Air Quality and Work Environment Survey: EPA Headquarters Building, Volume II: Results of Indoor Air Environmental Monitoring Study*, U.S. Environmental Protection Agency, Washington, DC, May 1990.

Environmental Protection Agency (EPA). 1990b. *Indoor Air Quality and Work Environment Survey: EPA Headquarters Buildings, Volume III: Relating Employee Responses to the Follow-Up Questionnaire with Environmental Measurements of Indoor Air Quality*, U.S. Environmental Protection Agency, Washington, DC, May 1990.

Environmental Protection Agency (EPA). 1990c. *Nonoccupational Pesticide Exposure Study*, Final Report EPA/600-S3-90-003. U.S. Environmental Protection Agency, Washington, DC.

Environmental Protection Agency (EPA). 1990d. *Nonoccupational Pesticide Exposure Study (NOPES) Project Summary*. U.S. Environmental Protection Agency, Research Triangle Park, NC.

Environmental Protection Agency (EPA). 1991. *Indoor Air Quality and Work Environment Survey: EPA Headquarters Buildings, Volume IV: Multivariate Statistical Analysis of Health, Comfort and Odor Perception as Related to Personal and Workplace Characteristics*, U.S. Environmental Protection Agency, Washington, DC, 1991.

Fischer, M.L., Bentley, A.J., et al. 1996. Factors affecting indoor air concentrations of volatile organic compounds at a site of subsurface gasoline contamination, *Environmental Science and Technology* 30(10):2948.

Girman, J.R., Hodgson, A.T. 1987. Exposure to methylene chloride from controlled use of a paint remover in a residence, presented at *80th Annual Meeting of the Air Pollution Control Association*, New York, NY, June 21–26, 1987, Report #LBL 23078, Lawrence Berkeley Laboratory, Berkeley, CA,.

Girman, J.R., Hodgson, A.T., et al. 1986. Volatile organic emissions from adhesives with indoor applications, *Environment International* 12(1–4):317–321.

Girman, J.R., Hodgson, A.T., et al. 1987. Considerations in evaluating emissions from consumer products, *Atmospheric Environment* 21:315–320.

Goldstein, B.D., Tardiff, R.G., et al. 1992. *Valdez Air Health Study*, Alyeska Pipeline Service Co., Anchorage, Alaska.

Goldstein, B.D., et al. 1979. The relationship between respiratory illness in primary school children and the use of gas for cooking: II. Factors affecting nitrogen dioxide levels in the home, *International Journal of Epidemiology* 8(4):339–345.

Guerin, M.R., Higgins, C.E., et al. 1987. Measuring environmental emissions from tobacco combustion: Sidestream cigarette smoke literature review, *Atmospheric Environment* 21:291–7.

Guerin, M.R., Jenkins, R.A., et al. 1992. *The Chemistry of Environmental Tobacco Smoke: Composition and Measurement*, Indoor Air Research Series, Center for Indoor Air Research, Lewis Publishers, Boca Raton, FL.

Health Effects Institute (HEI). 1995. *Particulate Air Pollution and Daily Mortality: Replication and Validation of Selected Studies, The Phase I Report of the Particle Epidemiology Evaluation Project.* Health Effects Institute, Cambridge, MA.

Heavner, D.L., Morgan, W.T., et al. 1995. Determination of volatile organic compounds and ETS apportionment in 49 homes, *Environment International* 21:3–21.

Higgins, C. E. 1987. Organic vapor phase composition of sidestream and environmental tobacco smoke from cigarettes, In: *Proceedings of the 1987 EPA/APCA Symposium on Measurement of Toxic and Related Air Pollutants*. 140–151.

Higgins, C.E. et al. 1983. Applications of Tenax trapping to cigarette smoking, *Journal-Association of Official Analytical Chemists* 66:1074–83.

Highsmith, V.R., Zweidinger, R.B., et al. 1988. Characterization of indoor and outdoor air associated with residences using woodstoves: A pilot study, *Environment International* 14:213–9.

Hodgson, A.T. et al. 1988. *Evaluation of Soil-Gas Transport of Organic Chemicals into Residential Buildings*: Report LBL-25465, Lawrence Berkeley Lab, Berkeley, CA.

Hodgson, A.T., Girman, J.R., et al. 1986. A multisorbent sampler for volatile organic compounds in indoor air, Paper No. 86-37.1, Presented at the *79th Annual Meeting of the Air Pollution Control Association*, Minneapolis, MN, June, 1986.

Hollowell, C. D., Miksch, R. R. 1981. Sources and concentrations of organic compounds in indoor environments. *Bulletin of the New York Academy of Medicine*, 57(10):962–977.

Jarke, F.H., Gordon, S.M. 1981. Recent investigations of volatile organics in indoor air at sub-ppb levels, Paper No. 81-57.2, Presented at the 74th Annual Meeting of the Air Pollution Control Association, Philadelphia, PA, June, 1981.

Jenkins, R.A., Guerin, M.R. 1994. Determination of Human Exposure to Environmental Tobacco Smoke, Interim Report No. 3, Center for Indoor Air Research, Linthicum, MD.

Jenkins, R.A., Palausky, A., et al. 1996. Exposure to environmental tobacco smoke in sixteen cities in the United States as determined by personal breathing zone air sampling, *Journal of Exposure Analysis and Environmental Epidemiology* 6(4):473–502.

Jenkins, R.A., Palausky, M.A., et al. 1995a. Determination of personal exposure of nonsmokers to environmental tobacco smoke in the United States, submitted to *Lung Cancer.*

Jenkins, R.A., Palausky, M.A., et al. 1995b. Personal exposure to environmental tobacco smoke in workplace and away from work settings: A 16 city case study, Presented at *Symposium on Measurement of Toxic and Related Air Pollutants*, May 17, 1995, Sponsored by the Air and Waste Management Association and the U.S. Environmental Protection Agency.

Jermini, C., Weber, A., et al. 1976. Quantitative determination of various gas-phase components of the sidestream smoke of cigarettes in room air (In German), *International Archives of Occupational and Environmental Health* 36:169–81.

Johansson, I. 1978. Determination of organic compounds in indoor air with potential reference to air quality, *Atmospheric Environment* 12:1371–1377.

Knöppel, H., Schauenburg, H. 1987. Screening of household products for the emission of volatile organic compounds, In: *Proceedings of the 4th International Conference on Indoor Air Quality and Climate*, Vol. 1, Institute for Soil, Water, and Air Hygiene, Berlin, 27–31

Kostiainen, R. 1995. Volatile organic compounds in the indoor air of normal and sick houses, *Atmospheric Environment* 29(6):693–702.

Koutrakis, P., Briggs, S.L.K., et al. 1992. Source apportionment of indoor aerosols in Suffolk and Onondaga Counties, New York, *Environmental Science and Technology* 26:521–27.

Krause, C., Mailahn, W., et al. 1987. Occurrence of volatile organic compounds in the air of 500 homes in the Federal Republic of Germany, In: *Proceedings of the 4th International Conference on Indoor Air Quality and Climate, Vol. 1,* Institute for Soil, Water, and Air Hygiene, Berlin, 102–106.

Krost, K.J., Pellizzari, E.D., et al. 1982. Collection and analysis of hazardous organic emissions, *Analytical Chemistry* 54:810–17.

Leaderer, B.P., Cain, W.S., et al. 1984. Ventilation requirements in buildings. II. Particulate matter and carbon monoxide from cigarette smoking, *Atmospheric Environment* 18: 99–106.

Leaderer, B.P., Hammond S.K. 1991. Evaluation of vapor-phase nicotine and respirable suspended particle mass as markers for environmental tobacco smoke, *Environmental Science and Technology* 25:770–777.

Lebret, E., Van de Weil, H.J., et al. 1986. Volatile hydrocarbons in Dutch homes, *Environment International* 12 (1–4):323–332.

Lewis, R.G., MacLeod, K.E. 1982. A portable sampler for pesticides and semivolatile industrial organic chemicals in air, *Analytical Chemistry* 54:310–315.

Lewis, R.G., Mulik, J.D., et al. 1985. Thermally desorbable passive sampling device for volatile organic chemicals in ambient air, *Analytical Chemistry* 57:214–219.

Lioy, P.J., Wallace, L.A., et al. 1991. Indoor/outdoor and personal monitor and breath analysis relationships for selected volatile organic compounds measured at three homes during New Jersey TEAM—1987, *Journal of Exposure Analysis and Environmental Epidemiology* 1(1):45–61.

Löfroth, G., Burton, B., et al. 1989. Characterization of environmental tobacco smoke, *Environmental Science and Technology* 23:610–614.

Löfroth, G., Stensman, C., et al. 1991. Indoor sources of mutagenic aerosol particulate matter: smoking, cooking, and incense burning, *Mutation Research* 261: 21–28.

Mahanama, K.R.R., Daisey, J.M. 1996. Volatile *n*-nitrosamines in environmental tobacco smoke: Sampling, analysis, emission factors, and indoor air exposures, *Environmental Science and Technology* 30(5):1477–1484.

Mailahn, W., Seifert, B., et al. 1987. The use of a passive sampler for the simultaneous determination of long-term ventilation rates and VOC concentrations, In: *Proceedings of the 4th International Conference on Indoor Air Quality and Climate,* Vol. 1, Institute for Soil, Water, and Air Hygiene, Berlin, 149–153

Marple, V.A., Rubow, K.L., et al. 1987. Low flow rate sharp cut impactors for indoor air sampling: Design and calibration, *Journal of the Air Pollution Control Association* 37:1303–1307.

Miksch, R.R., Hollowell, C.D., et al. 1982. Trace organic chemical contaminants in office spaces, *Environment International* 8:129–137.

Mølhave, L. 1982. Indoor air pollution due to organic gases and vapours of solvents in building materials, *Environment International* 8(1–6):117–127.

Mølhave, L., Møller, J. 1979. The atmospheric environment in modern Danish dwellings: Measurements in 39 flats, In: *Indoor Climate,* Fänger, O., Valbjörn, Eds., SBI: Hörsholm, Denmark, 171–186.

National Institute for Occupational Safety and Health (NIOSH). 1991a. *Indoor Air Quality and Work Environment Study: Volume I: Results of Employee Survey,* Cincinnati, OH, HETA 88-364-2102.

National Institute for Occupational Safety and Health (NIOSH). 1991b. *Indoor Air Quality and Work Environment Study: Volume II: Results of Indoor Air Environmental Monitoring,* Cincinnati, OH, HETA 88-364-2103.

National Institute for Occupational Safety and Health (NIOSH). 1991c. *Indoor Air Quality and Work Environment Study: Volume III: Association Between Health and Comfort Concerns and Environmental Conditions,* Cincinnati, OH, HETA 88-364-2104.

National Toxicology Program (NTP). 1986. *Technical Report on the Toxicity and Carcinogenesis of 1,4-Dichlorobenzene (CAS #106-46-7) in F344/n Rats and B6C3F1 Mice (Gavage Study)*. National Toxicology Program Technical Report # 319, Board Draft, March 1986.

National Toxicology Program (NTP). 1988. *Technical Report on the Toxicity and Carcinogenesis of d-Limonene (CAS #5989-27-5) in F344/N Rats and B6C3F1 Mice (Gavage Study)*. National Toxicology Program Technical Report # 347, NIH Publication # 88-2802.

Nazaroff, W.W., Cass, G.R. 1989. Mathematical modeling of indoor aerosol dynamics, *Environmental Science and Technology* 23:157–166.

Nazaroff, W.W., Gadgil, A.J., et al. 1993. Critique of the use of deposition velocity in modeling indoor air quality, In: Nagda, N.L. Ed., *Modeling of Indoor Air Quality and Exposure*, ASTM STP 1205, American Society for Testing and Materials: Philadelphia, 81–104.

Nazaroff, W.W., Ligocki, M.P., et al. 1990b. Particle deposition in museums: Comparison of modeling and measurement results, *Aerosol Sci. Tech.*, 13:332–348.

Nazaroff, W.W., Salmon, L.G., et al. 1990a. Concentration and fate of airborne particles in museums, *Environmental Science and Technology* 24:66–77.

Neas, L.M., Dockery, D.W., et al. 1994. Concentration of indoor particulate matter as a determinant of respiratory health in children, *American Journal of Epidemiology* 139:1088–1099.

Nuchia, E. 1986. MDAC — Houston Materials Testing Database Users' Guide.

Offermann, F.J., Sextro, R.G., et al. 1985. *Atmospheric Environment* 19:1761–1771.

Ogden, M.W., Heavner D.L., et al. 1995. Personal monitoring system for measuring environmental tobacco smoke exposure, Accepted by *Environmental Technology*.

Ogden, M.W., Maiolo, K.C. 1989. Collection and determination of solanesol as a tracer of environmental tobacco smoke in indoor air, *Environmental Science and Technology*, 23:1148–1154.

Ogden, M.W., Maiolo, K.C., et al. 1990. Evaluation of methods for estimating the contribution of ETS to respirable suspended particles, In: *Indoor Air '90: Proceedings of the 5th International Conference on Indoor Air Quality and Climate*, Vol. 4, Walkinshaw, D.S., Ed., Canada Mortgage and Housing Corp.: Ottawa, Ontario, 415–420.

Oliver, K.D., Pleil, J.D., et al. 1986. Sample integrity of trace level volatile organic compounds in ambient air stored in summa polished canisters, *Atmospheric Environment* 20:1403.

Ott, W., Thomas, J., et al. 1988. Validation of the simulation of human activity and agent exposure (SHAPE) model using paired days from the Denver, CO, carbon monoxide study, *Atmospheric Environment* 22:2101–2113.

Özkaynak, H., Ryan, P.B., et al. 1987. Sources and emission rates of organic chemical vapors in homes and buildings, In: *Proceedings of the 4th International Conference on Indoor Air Quality and Climate, Vol. 1*, Institute for Soil, Water, and Air Hygiene: Berlin, 3–7.

Özkaynak, H., Spengler, J.D., et al. 1993. Sources and factors influencing personal and indoor exposures to particles: Findings from the particle TEAM pilot study, In: *Indoor Air '93: Proceedings of the 6th International Conference on Indoor Air Quality and Climate, Vol. 3*:457–462.

Özkaynak, H., Xue, J., et al. 1996a. *The Particle TEAM (PTEAM) Study: Analysis of the Data*, Vol. III, *Final Report*, Contract #68-02-4544, U.S. Environmental Protection Agency, Research Triangle Park, NC.

Özkaynak, H., Xue, J., et al. 1996b. Personal exposure to airborne particles and metals: Results from the particle TEAM study in Riverside, CA, *Journal of Exposure Analysis and Environmental Epidemiology* 6:57–78.

Palmes, E.D., Gunnison, A.F., et al. 1986. Personal sampler for NO_2. *American Industrial Hygiene Association Journal* 37:570–577.

Pellizzari, E.D., Perritt, K., et al. 1987a. *Total Exposure Assessment Methodology (TEAM) Study: Elizabeth and Bayonne, New Jersey; Devils Lake, North Dakota, and Greensboro, North Carolina,* Vol. II, U.S. Environmental Protection Agency, Washington, DC.

Pellizzari, E.D., Perritt, K., et al. 1987b. *Total Exposure Assessment Methodology (TEAM) Study: Selected Communities in Northern and Southern California,* Vol. III, U.S. Environmental Protection Agency, Washington, DC.

Pellizzari, E.D., Thomas, K.W., et al. 1992. *Particle Total Exposure Assessment Methodology (PTEAM): Riverside, California Pilot Study,* Final Report, Vol. 1, EPA Contract # 68-02-4544.,U.S. Environmental Protection Agency, Research Triangle Park, NC.

Phillips, K., Bentley, M.C., et al. 1996. Assessment of air quality in Stockholm by personal monitoring of nonsmokers for respirable suspended particles and environmental tobacco smoke, *Scandinavian Journal of Work, Environment and Health* 22 (Suppl.):1–24.

Phillips, K., Bentley, M.C., et al. 1997a. Assessment of air quality in Barcelona by personal monitoring of nonsmokers for respirable suspended particles and environmental smoke, *Environment International* 23(2):173–196.

Phillips, K., Howard, D.A., et al. 1994. Assessment of personal exposures to environmental tobacco smoke in British nonsmokers, *Environment International* 20(6):693–712.

Phillips, K., Howard, D.A., et al. 1997b. Assessment of air quality in Turin by personal monitoring of nonsmokers for respirable particles and environmental tobacco smoke, *Environment International* 23(6):851–871.

Raunemaa, T., Kulmala, M., et al. 1989. Indoor air aerosol model: Transport indoors and deposition of fine and coarse particles, *Aerosol Sci. Tech.* 11:11–25.

Repace, J.L., Lowrey, A.H. 1980. Indoor air pollution, tobacco smoke, and public health, *Science* 208:464–472.

Repace, J.L., Lowrey, A.H. 1982. Tobacco smoke, ventilation, and indoor air quality.*ASHRAE Transactions* 88(1):895–914.

Repace, J.L., Lowrey, A.H. 1985. A quantitative estimate of non-smokers' lung cancer risk from passive smoking, *Environment International* 11:3–22.

Rupprecht, E., Meyer, M., et al. 1992. The tapered element oscillating microbalance as a tool for measuring ambient particulate concentrations in real time, *J. Aersol. Sci.* 23(suppl. 1):S635–S638.

Ryan, P.B., Schwab, M., et al. 1992. *Nitrogen Dioxide Exposure Studies—Volume III: Personal Exposure to Nitrogen Dioxide in Boston: A Microenvironmental Approach,* Gas Research Institute: Chicago, IL.

Ryan, P. B., Soczek, M.L., et al. 1988a. The Boston residential NO_2 characterization study: I. A preliminary evaluation of the survey methodology, *Journal of the Air Pollution Control Association* 38:22–27.

Ryan, P.B., Soczek, M.L., et al. 1988b. The Boston residential NO_2 characterization study: II. Survey methodology and population concentration estimates, *Atmospheric Environment* 22:2115–25.

Santanam, S., Spengler, J.D., et al. 1990. Particulate matter exposures estimated from an indoor-outdoor source apportionment study, In: Walkinshaw, D.S., Ed., *Indoor Air '90: Proceedings of the 5th International Conference on Indoor Air Quality and Climate,* Vol. 2, Canada Mortgage and Housing Corp.: Ottawa, Ontario, 583–588.

Schwab, M., Spengler, J.D., et al. 1990. *Nitrogen Dioxide Exposure Studies, Vol. IV: Human Exposure to Nitrogen Dioxide in the Los Angeles Basin,* Gas Research Institute: Chicago, IL, and Southern California Gas Company: Los Angeles, CA.

Seifert, B., Abraham, H.J. 1982. Indoor air concentrations of benzene and some other aromatic hydrocarbons, *Ecotoxicology and Environmental Safety* 6:190–192.

Seifert, B., Abraham, H.J. 1983. Use of passive samplers for the determination of gaseous organic substances in indoor air at low concentration levels, *International Journal of Environmental Analytical Chemistry* 13:237–253.

Seifert, B., Schulz, C., et al. 1987. Seasonal variation of concentrations of volatile organic compounds in selected German homes, In: *Proceedings of the 4th International Conference on Indoor Air Quality and Climate,* Vol. 1, Institute for Soil, Water, and Air Hygiene: Berlin, 107–111.

Shair, F.H., Heitner, K.L. 1974. Theoretical model for relating indoor pollutant concentrations to those outside, *Environmental Science and Technology* 8:444–451.

Sheldon, L.S., Clayton, A., et al. 1993. Indoor concentrations of polycyclic aromatic hydrocarbons in California residences and their relationship to combustion source use, In: *Indoor Air '93: Proceedings of the 6th International Conference on Indoor Air Quality and Climate,* Vol. 3, Helsinki University of Technology: Espoo, Finland, 29–34.

Sheldon, L.S., Hartwell, T.D., et al. 1989. *An Investigation of Infiltration and Indoor Air Quality, Final Report,* Contract # 736-CON-BCS-85, New York State Energy Research and Development Authority, Albany, NY.

Sinclair, J.D., Psota-Kelty, L.A., et al. 1988., *Atmospheric Environment* 22:461–469.

Sinclair, J.D., Psota-Kelty, L.A., et al. 1990. Measurement and modelling of airborne concentrations and indoor surface accumulation rates of ionic substances at Neenah, Wisconsin, *Atmospheric Environment* 24A:627–638.

Sinclair, J.D., Psota-Kelty, L.A., et al. 1992. Indoor-outdoor relationships of airborne ionic substances: Comparison of electronic equipment room and factory environments, *Atmospheric Environment* 26A:871–882.

Skov, P., Valbjorn, O. 1987. The "sick" building syndrome in the office environment: The Danish town hall study, In: *Indoor Air '87: Proceedings of the 4th International Conference on Indoor Air Quality and Climate, August 17–21, 1987, Vol. 2,* Institute for Water, Soil, and Air Hygiene: Berlin, 439–443.

Skov, P., Valbjorn, O., et al. 1989. Influence of personal characteristics, job-related factors, and stress factors on the sick building syndrome, *Scandinavian Journal Work Environment Health* 15:286–295.

Skov, P., Valbjorn, O., et al. 1990. Influence of indoor climate on the sick building syndrome in an office environment, *Scandinavian Journal of Work Environment Health* 16:363–371.

Sparks, L.E., Tichenor, B.A., et al. 1993. Modeling individual exposure from indoor sources, In: *Modeling of Indoor Air Quality and Exposure,* Nagda, N.L., Ed., ASTM: Philadelphia, PA, ASTM STP 1205.

Sparks, L.E., Tichenor, B.A., et al. 1991. Comparison of data from an IAQ test house with predictions of an IAQ computer model, *Indoor Air* 1:577–592.

Spengler, J.D., Dockery, D.W., et al. 1980. Long-term measurements of respirable sulfates and particles inside and outside homes, *Atmospheric Environment* 15:23–30.

Spengler, J.D., Dockery, D.W., et al. 1981. Long-term measurements of respirable sulfates and particles inside and outside homes, *Atmospheric Environment* 15:23–30.

Spengler, J.D., Duffy, C.P., et al. 1983. Nitrogen dioxide concentrations inside and outside 137 homes and implications for ambient air quality standards and health effects research, *Environmental Science and Technology* 17:164–168.

Spengler, J.D., Ryan, P.B., et al. 1992. *Nitrogen Dioxide Exposure Studies—Vol. 4, Personal Exposure to Nitrogen Dioxide in the Los Angeles Basin.* Gas Research Institute: Chicago, IL.

Spengler, J.D., Thurston, G.D. 1983. Mass and elemental composition of fine and coarse particles in six U.S. cities, *Journal of the Air Pollution Control Association* 33:1162–1171.

Spengler, J.D., Treitman, R.D., et al. 1985. Personal exposures to respirable particulates and implications for air pollution epidemiology, *Environmental Science and Technology* 19:700–707.

Spengler, J.D., Ware, J., et al. 1987. Harvard's Indoor Air Quality Respiratory Health Study, In: *Indoor Air '87: Proceedings of the 4th International Conference on Indoor Air Quality and Climate,* August 17–21, 1987, Vol. 2, Seifert, B., Ed., Institute for Water, Soil, and Air Hygiene: Berlin, 742–746.

Spicer, C.W., et al. 1986. Intercomparison of sampling techniques for toxic organic compounds in indoor air, In: *Proceedings of the 1986 EPA/APCA Symposium on the Measurement of Toxic Air Pollutants,* Hochheiser, S. and Jayanti, R.K.M., Eds., Air Pollution Control Association: Pittsburgh, PA, 45–60.

Thomas, K.W., Pellizzari, E.D., et al. 1991. Effect of dry-cleaned clothes on tetrachloroethylene levels in indoor air, personal air, and breath for residents of several New Jersey homes, *Journal of Exposure Analysis and Environmental Epidemiology* 1:475–490.

Thomas, K.W., Pellizzari, E.D., et al. 1993a. Particle Total Exposure Assessment Methodology (PTEAM) study: Method performance and data quality for personal, indoor, and outdoor aerosol monitoring at 178 homes in Southern California, *Journal of Exposure Analysis and Environmental Epidemiology* 3:203–226.

Thomas, K.W., Pellizzari, E.D., et al. 1993b. Temporal variability of benzene exposure for residents in several New Jersey homes with attached garages or tobacco smoke, *Journal of Exposure Analysis and Environmental Epidemiology* 3:49–73.

Tichenor, B.A., Mason, M.A. 1987. Organic emissions from consumer products and building materials to the indoor environment, *Journal of the Air Pollution Control Association* 38:264–68.

Tichenor, B.A., Sparks, L.E., et al. 1990. Evaluating sources of indoor air pollution, *Journal of the Air and Waste Management Association* 41:487–92.

Traynor, G.W., Aceti, J.C., et al. 1989. *Macromodel for Assessing Residential Concentrations of Combustion-generated Pollutants: Model Development and Preliminary Predictions for CO, NO₂, and Respirable Suspended Particulates,* LBL-25211, Lawrence Berkeley Laboratory: Berkeley, CA.

Traynor, G.W., Apte, M.G., et al. 1990. Selected organic pollutant emissions from unvented kerosene space heaters, *Environmental Science and Technology* 24:1265–70.

Turk, B.H., Brown, J.T., et al. 1987. Indoor Air Quality and Ventilation Measurements in 38 Pacific Northwest Commercial Buildings, Volume I: Measurement Results and Interpretation, Final Report, Report LBL-22315, Lawrence Berkeley Laboratory.

Turk, B.H., Grimsrud, D.T., et al. 1989. Commercial building ventilation rates and particle concentrations, *ASHRAE Transactions* 95:422–433.

Wallace, L.A. 1987. *The TEAM Study: Summary and Analysis: Volume I,* EPA 600/6-87/002a, U.S.Environmental Protection Agency, Washington, DC.

Wallace, L.A. 1989. The exposure of the general population to benzene, *Cell Biology and Toxicology* 5:297–314.

Wallace, L.A. 1990. Major sources of exposure to benzene and other volatile organic compounds, *Risk Analysis* 10:59–64.

Wallace, J.C., Basu, I., et al. 1996. Sampling and analysis artifacts caused by elevated indoor air polychlorinated biphenyl concentrations, *Environmental Science and Technology* 30(9):2730–2734.

Wallace, L.A., Nelson, W.C., et al. 1991a. The Los Angeles TEAM study: Personal exposures, indoor-outdoor air concentrations, and breath concentrations of 25 volatile organic compounds, *Journal of Exposure Analysis and Environmental Epidemiology* 1(2):37–72.

Wallace, L.A., Pellizzari, E.,et al. 1984. Personal exposure to volatile organic compounds: I. Direct measurement in breathing-zone air, drinking water, food, and exhaled breath. *Environmental Research* 35:293–319.

Wallace, L.A., Pellizzari, E., et al. 1985. Personal exposures, indoor-outdoor relationships and breath levels of toxic air agents measured for 355 persons in New Jersey,*Atmospheric Environment* 19:1651–1661.

Wallace, L.A., Pellizzari, E., et al. 1986. Total Exposure Assessment Methodology (TEAM) study: Personal exposures, indoor-outdoor relationships, and breath levels of volatile organic compounds in New Jersey, *Environment International* 12:369–387.

Wallace, L.A., Pellizzari, E., et al. 1987a. Exposures to benzene and other volatile organic compounds from active and passive smoking, *Archives of Environmental Health* 2:272–279.

Wallace, L.A., Pellizzari, E., et al. 1987b. The TEAM study: Personal exposures to toxic substances in air, drinking water, and breath of 400 residents of New Jersey, North Carolina, and North Dakota, *Environmental Research* 43:290–307.

Wallace, L.A., Pellizzari, E., et al. 1987c. Emissions of volatile organic compounds from building materials and consumer products, *Atmospheric Environment* 21:385–393.

Wallace, L.A., Pellizzari, E.D., et al. 1988a. The California TEAM study: Breath concentrations and personal exposures to 26 volatile compounds in air and drinking water of 188 residents of Los Angeles, Antioch, and Pittsburgh, CA, *Atmospheric Environment* 22:2141–63.

Wallace, L.A., Pellizzari, E.D., et al. 1989. The influence of personal activities on exposure to volatile organic compounds, *Environmental Research* 50:37–55.

Wallace, L.A., Pellizzari, E., et al. 1991b. Total volatile organic concentrations in 2700 personal, indoor, and outdoor air samples collected in the U.S. EPA TEAM studies, *Indoor Air* 4:465–477.

Wallace, L.A., Thomas, J., et al. 1988b. Comparison of breath CO, CO exposure, and Coburn model predictions in the U.S. EPA Washington-Denver CO study,*Atmospheric Environment* 22:2183–93.

Wallace, L.A., Zweidinger, R., et al. 1982. Monitoring individual exposure: Measurement of volatile organic compounds in breathing-zone air, drinking water, and exhaled breath, *Environment International* 8:269–282.

Weschler, C.J. 1978. Characterization techniques applied to indoor dust, *Environmental Science and Technology* 12(8):923–926.

Weschler, C.J., Shields, H.C., et al. 1989. Comparison of effects of ventilation, filtration and outdoor air at telephone office buildings, In: *Design and Protocol for Monitoring Indoor Air Quality, ASTM STP 1002*, Nagda N.L., Harper, J.P., Eds, American Society for Testing and Materials: Philadelphia. PA, 19–34.

Wilson, A.L., Colome, S.D., et al. 1986. Residential Indoor Air Quality Characterization Study of Nitrogen Dioxide, Phase I Final Report, Southern California Gas Company: Los Angeles, CA.

Wilson, A.L., Colome, S.D., et al. 1993a. *California Residential Indoor Air Quality Study, Volume I: Methodology and Descriptive Statistics*, Integrated Environmental Services: Irvine, CA.

Wilson, A.L., Colome, S.D., et al. 1993b. *California Residential Indoor Air Quality Study, Volume I: Appendix*, Integrated Environmental Services: Irvine, CA.

Wilson, A.L., Colome, S.D., et al. 1995a. *California Residential Indoor Air Quality Study, Volume III: Ancillary and Exploratory Analyses,* Integrated Environmental Services: Irvine, CA.

Wilson, N.K., Barbour, R.K., et al. 1994. Evaluation of a real-time monitor for fine particle-bound PAH in air, *Polycyclic Aromatic Compounds* 5:167-174.

Yanagisawa, Y., Nishimura, H. 1982. A badge-type personal sampler for measurement of personal exposure to NO_2 and NO in ambient air, *Environment International* 8:235–239.

ADDITIONAL BIBLIOGRAPHIC MATERIAL

Berglund, B., Berglund, U., et al. 1984. Characterization of indoor air quality and "sick buildings," *ASHRAE Transactions* 90(Pt. 1): 1045–1055.

Bridge, D. P., Corn, M. 1972. Contribution to the assessment of exposure of nonsmokers to air pollution from cigarette and cigar smoke in occupied spaces, *Environmental Research* 5: 192–209.

Brief, R. S. 1960. A simple way to determine air contaminants, *Air Engineering* 2:39–51.

Environmental Protection Agency (EPA). 1996. *Exposure Factors Handbook, Vol. III: Activity Factors,* Washington, DC, EPA/600/P-95/002Bc.

Gordon, S.M., Kenny, D.V., et al. 1992. Continuous real-time breath analysis for the measurement of half-lives of expired volatile organic compounds, *Journal of Exposure Analysis and Environmental Epidemiology,* Suppl. 1:41–54.

Ishizu, Y. 1980. General equation for the estimation of indoor pollution, *Environmental Science and Technology* 14(10):1254–1257.

Jones, R.M., Fagan, R. 1974. Application of mathematical model for the buildup of carbon monoxide from cigarette smoking in rooms and houses, *ASHRAE Journal* 16:49–53.

Klepeis, N.E., Ott, W.R., et al. 1995. Modeling the time series of respirable suspended particles and carbon monoxide from multiple smokers: Validation in two public smoking lounges. Paper # A-1233, presented at the *88th Annual Meeting of the Air and Waste Management Association.*

Lewis, R.G., Bond, A.E. 1987. Nonoccupational exposure to household pesticides, *Proceedings of The 4th International Conference on Indoor Air Quality and Climate,* Vol. 1, Institute for Soil, Water, and Air Hygiene: Berlin, 195–9.

Ott, W.R. 1985. Total human exposure: An emerging science focuses on humans as receptors of environmental pollution, *Environmental Science Technology* 19:880.

Ott, W.R. 1990a. Total human exposure: Basic concepts, EPA field studies, and future research needs, *Journal of the Air and Waste Management Association* 40(7):966–975.

Ott, W.R. 1990b. A physical explanation of the lognormality of agent concentrations, *Journal of the Air and Waste Management Association* 40:1378–83.

Ott, W.R., Klepeis, N.E., et al. 1995b. Modeling environmental tobacco smoke in the home using transfer functions, Paper number A1043 presented at the *88th Annual Meeting of the Air and Waste Management Association.*

Ott, W.R., Langan, L., et al. 1992. A time series model for cigarette smoking activity patterns: Model validation for carbon monoxide and respirable particles in a chamber and an automobile, *Journal of Exposure Analysis and Environmental Epidemiology* 2(2): 175–200.

Ott, W.R., Switzer, P., et al. 1995a. Particle concentrations inside a tavern before and after prohibition of smoking: Evaluating the performance of an indoor air quality model, submitted for publication.

Pellizzari, E.D., Michael, L.C., et al. 1988. *Comparison of Indoor and Outdoor Toxic Air Agent Levels in Several Southern California Communities*, Final Report, Contract # 68-02-4544. U.S. Environmental Protection Agency: Research Triangle Park, NC.

Pellizzari, E.D., Thomas, K.W., et al. 1990. *Measurements of Exhaled Breath Using a New Portable Sampling Method*, Final Report. Contract # 68-02-4544, U.S. Environmental Protection Agency: Research Triangle Park, NC.

Repace, J.L. 1987a. Indoor concentrations of environmental tobacco smoke: Models dealing with effects of ventilation and room size, Volume 9, Passive Smoking, In: *Environmental Carcinogens—Methods of Analysis and Exposure Measurement,* O'Neill, I.K., Brunnemann, K.D., Dodet, B., and Hoffmann, D., Eds., International Agency for Research on Cancer: Lyon, France, 25–41.

Repace, J.L. 1987b. Indoor concentrations of environmental tobacco smoke: Field surveys, Volume 9, Passive Smoking, in *Environmental Carcinogens—Methods of Analysis and Exposure Measurement,* O'Neill, I.K., Brunnemann, K.D., Dodet, B., and Hoffmann, D., Eds., International Agency for Research on Cancer: Lyon, France, 141–162.

Roberts, J.W., Budd, W.T., et al. 1992. Human exposure to pollutants in the floor dust of homes and offices, *Journal of Exposure Analysis Environmental Epidemiology,* Suppl. 1:127–146.

Roberts, J.W., Ruby, M.G., et al. 1987. Mutagenic activity of house dust, In: *Short-Term Bioassays in the Analysis of Complex Environmental Mixtures,* Sandhu, S. S., De Marini, D. M., Mass, M. J., Moore, M. M., and Mumford, J. L., Eds., V. Plenum Press: New York, 355–367.

Sigsby, J.E., Tejada, S.B., et al. 1987. Volatile organic compound emissions from 46 in-use passenger cars, *Environmental Science and Technology* 21:466–475.

South Coast Air Quality Management District (SCAQMD). 1989. *In-Vehicle Characterization Study in the South Coast Air Basin,* SCAQMD: Los Angeles, CA.

Spengler, J.D., Özkaynak, H., et al. 1989. Personal exposures to particulate matter: Instruments and methodologies for PTEAM study, presented at *EPA-AWMA Conference on Total Exposure Assessment Methodology—A New Horizon,* Las Vegas, NV.

Thatcher, T.L., Layton, D.W. 1995. Deposition, resuspension, and penetration of particles within a residence, *Atmospheric Environment* 29:1487–1497.

Turk, A. 1963. Measurements of odorous vapors in test chambers: Theoretical, *ASHRAE Journal* (5)10: 55–58.

Wallace, L.A. 1991. Comparison of risks from outdoor and indoor exposure to toxic chemicals, *Environmental Health Perspectives* 95:7–13.

Wallace, L.A. 1993. A decade of studies of human exposure: what have we learned? *Risk Analysis* 13:135–139.

Wallace, L.A. 1995. Human exposure to environmental agents: A decade of experience, *Clinical and Experimental Allergy* 25:4–9.

Wallace, L.A. 1996. Indoor particles: A review, *JAWMA* 46:98–126.

Wallace, L.A. 1997. Human exposure and body burden for chloroform and other trihalomethanes, *Critical Reviews in Environ. Sci. Tech.* 27:113–194.

Wallace, L.A., Buckley, T., et al. 1996. Breath measurements as VOC biomarkers: EPA's experience in field and chamber studies, *Environmental Health Perspectives* 104(Suppl. 5): 861–869.

Wallace, L.A., Nelson, C.J., et al. 1993. Association of personal and workplace characteristics with health, comfort, and odor: A survey of 3948 office workers in three buildings, *Indoor Air* 3:193–205.

Wallace, L.A., Nelson, C.J., et al. 1995. Perception of indoor air quality among government employees in Washington, DC., *Technology: J. Franklin Institute* 332A:183–198.

Wallace, L. A., Nelson, W.C., et al. 1997. A four compartment model relating breath concentrations to low-level chemical exposures: Application to a chamber study of five subjects exposed to nine VOCs, *Journal of Exposure Analysis and Environmental Epidemiology* 7(2):141–163.

Zweidinger, R.B., Sigsby, J.E., et al. 1988. Detailed hydrocarbon and aldehyde mobile source emissions from roadway studies, *Environmental Science and Technology* 22:956–62.

CHAPTER 9

Application of Risk Assessment

David R. Patrick

CONTENTS

1-56670-323-9/99/$0.00+$.50
© 1999 by CRC Press LLC

I. INTRODUCTION

This chapter provides an example of how a risk assessment is applied to a specific substance in a specific setting. The intent is to take a real world environmental risk and present each of the four steps of a risk assessment separately to show how each functions in the estimation of risks to exposed humans. Risk management options are also discussed. Because most pollutants that can be readily assessed can exist both outdoors and indoors, the example selected is of concern to both environments.

The air pollutant chosen for this example is particulate matter (PM). PM is chosen in large part because it is ubiquitous and because there are substantial scientific controversies over the health effects resulting from low-level exposures occurring indoors and outdoors. As such, the reader can readily see how uncertainties in risk assessments arise and are treated.

PM is a broad class of chemically and physically diverse substances that exist as discrete particles of condensed liquid or solid materials. PM can exist in a wide range of sizes, from molecular clusters 0.005 microns in diameter to coarse particles on the order of 100 microns. PM also can exist in a wide range of compositions including elements, inorganic compounds, organic compounds, and mixtures of the preceding. Importantly to human health, particles smaller than about 10 microns in diameter are thought to be of more health concern because larger particles are not taken as deeply into the lung. Recent research also shows that particles below a few microns in size can reach even more deeply into the lung than 10 micron particles and may result in more serious adverse effects, although there is considerable uncertainty about the effective size. However, larger particles can also represent a concern for some adverse health effects when they are deposited in the nasal and mucous membranes and then ingested, and when contacted by the skin and subsequently absorbed or ingested.

PM is a health concern both outdoors and indoors. Significant outdoor sources of PM include fuel combustion (e.g., vehicles, power generation, and industrial facilities), residential fireplaces, agricultural and forest burning, atmospheric formation from gaseous precursors (largely produced from fuel combustion), and wind-blown dust. Significant indoor sources of PM include fuel combustion (e.g., heating and cooking), tobacco smoke, cleaning practices, and infiltration of outdoor air. Outdoor PM is regulated by the EPA and state and local air pollutant control agencies. Indoor PM is not federally regulated except for workplace standards for specific substances that are established and enforced by the U.S. Occupational Safety and Health Administration (OSHA). Before summarizing available information regarding the potential risks resulting from exposure to PM, the EPA and the OSHA regulatory processes are briefly described and the current regulations are summarized.

The appropriate regulation of PM was the source of great controversy in the mid-1990s. Following a lengthy and heated debate, the EPA promulgated revisions to the outdoor air PM standards on July 18, 1997 (62 FR 38652). At the time that this book was written, the debate on the standards continued and members of Congress were threatening to delay or repeal the standards. Much of the information here is summarized from the extensive and complex record of that regulatory action.

However, to facilitate the use of this book by a broad range of readers, that record is only summarized here and only the major references are cited. Detailed discussions of the underlying science and the controversies are better obtained from the original sources. The key EPA references used to prepare this chapter were the Criteria Document (EPA 1996a) and the Staff Paper (EPA 1996b). All documents relevant to the promulgated PM standards can be found in the EPA regulatory docket.

II. FEDERAL REGULATION OF PARTICULATE MATTER

A. The Regulatory Processes

1. Outdoor Particulate Matter

PM is regulated by the EPA as a criteria air pollutant. Criteria pollutants are defined as pollutants whose sources are numerous and diverse. They were originally assumed to be pollutants for which a safe level of exposure could be established, although more recently this assumption is being challenged in certain cases. The 1970 Amendments to the Clean Air Act (CAA) initially established the process for regulating these pollutants. Section 108 required the EPA to identify air pollutants that "may reasonably be anticipated to endanger the public health and welfare." For such pollutants, the EPA was to issue air quality criteria in a Criteria Document, hence the term "criteria pollutant." Section 109 then required the EPA to propose and promulgate primary and secondary National Ambient Air Quality Standards (NAAQS) based on the air quality criteria. A primary NAAQS must protect the public health with an "adequate margin of safety,"[1] while a secondary NAAQS must protect the public welfare[2] from any "known or anticipated effects."

The requirement to protect the public health with an adequate margin of safety was intended to account for uncertainties arising from incomplete scientific information and to provide reasonable protection against hazards not yet identified. The NAAQS process for selecting primary standards has been interpreted by the EPA and the Courts as a health-based decision process that excludes consideration of costs and other impacts. Costs and other impacts are to be considered only in the strategies for complying with the NAAQS. The EPA and the Courts interpret the CAA as not requiring NAAQS to be set a "zero risk" level.

Section 109 further required the EPA to review and, if appropriate, revise the NAAQS every 5 years. It also required the appointment of "an independent scientific review committee composed of seven members, [initially] including one member from the National Academy of Sciences, one physician, and one person representing State air pollution control agencies." This Committee is called the Clean Air Scientific Advisory Committee (CASAC); it reviews and comments on the EPA NAAQS criteria document and the proposed regulatory actions.

---

[1] The legislative history of Section 109 states that primary standards are to be set at levels that protect the most sensitive group of the population rather than the average population.

[2] A welfare effect is any effect that is not a human health effect.

The regulatory process used by the EPA to revise a NAAQS usually takes longer than the 5 years required by the CAA. The process typically involves the following steps: (a) preparation of a comprehensive Criteria Document by the EPA that details the current knowledge on health and welfare effects; (b) review of the Criteria Document by the CASAC; (c) preparation of a detailed Staff Paper by the EPA that interprets the Criteria Document and suggests a range of possible standards for consideration; (d) review of the Staff Paper by the CASAC; (e) proposal of a regulation; (f) public review and comment; and (g) promulgation of a final standard.

As initially conceived, the EPA was to determine the safe level of exposure necessary to protect the most sensitive group of the population. Such groups might be children (who are often outdoors more frequently than adults and are more active), outdoor workers (who may be active), individuals with respiratory diseases (including asthma, emphysema, and chronic obstructive pulmonary disease), and otherwise healthy individuals who are especially sensitive to the pollutant of concern. In the early days, before the science of risk assessment began to mature, the regulatory decisions were made strictly based on this approach. More recently, broader potential impacts of exposures to a pollutant are used in deciding the final levels and types of standards. For example, the health effects evidence (e.g., human clinical, epidemiology, and animal toxicology) continues to be used in conjunction with information on the underlying uncertainties. However, these are being supplemented with broader information on "at risk" populations, the degree of human exposure to levels at which adverse effects are observed, the estimated size of populations at risk, and air quality comparisons across the air sampling monitor sites in areas where standards are met.

2. Indoor Particulate Matter

The OSHA regulates substances in the workplace air by establishing and enforcing Permissible Exposure Limits (PELs). These were authorized in Section 6 (Occupational Safety and Health Standards) of the Occupational Safety and Health Act, enacted in 1970. Section 6(b)(5) requires standards for toxic materials and harmful physical agents to be set at a level that "most adequately assures, to the extent feasible, on the basis of the best available evidence, that no employee will suffer material impairment of health or functional capacity even if such employee has regular exposure to the hazard dealt with by such standard for the period of his working life." The process used by the OSHA for setting PELs typically involves Advisory Committees that are called on to develop specific recommendations. There are two standing advisory committees, and ad hoc committees may be appointed to examine special areas of concern to the OSHA. All committees must have members representing management, labor, and state agencies. The two standing advisory committees are:

1. National Advisory Committee on Occupational Safety and Health, and
2. Advisory Committee on Construction Safety and Health.

Recommendations for standards can also come from the National Institute for Occupational Safety and Health (NIOSH), which was also formed as a result of the

1970 Occupational Safety and Health Act. NIOSH is an agency of the Department of Health and Human Services formed to conduct research on various safety and health problems, provide technical assistance to the OSHA, and recommend standards for OSHA adoption.

Once the need for a PEL for a specific substance is verified and recommendations are received from the appropriate Advisory Committee and NIOSH, the OSHA may publish an advance notice of proposed rulemaking in order to gather more data, or directly propose a standard. Following receipt and review of public comment, including a public hearing if requested, the OSHA promulgates a final standard.

While the OSHA safety standards require a cost balancing (Section 3[8] requires use of practices, means, methods, operations, or processes reasonably necessary or appropriate), health standards are not so constrained. The Courts have also interpreted Section 6(b)(5) as meaning that Congress has already made the cost-benefit calculation and required that standards err on the side of health protection. In addition, the requirements are viewed as technology forcing. However, the OSHA is required to determine that a risk exists, the degree to which the standard will reduce the risk, and the feasibility of the standard. Certain rules have been overturned by the Courts which judged that the OSHA had not met those requirements.

B. Current Particulate Matter Standards

1. Outdoor Particulate Matter

Human health effects resulting from exposures to air pollutants are usually assessed through methods involving statistical techniques. Because there is reasonable access today to detailed data on populations, exposures, and hospital records, epidemiological studies are widely used. However, studies of large populations are often necessary because pollutants in the ambient air usually exist at relatively low concentrations and the health effects resulting from exposure to these concentrations can be subtle. In addition, the U.S. population is highly diverse in genetic makeup, socioeconomic position, and lifestyle. Typical exposures can also vary significantly because the U.S. population is highly mobile and often moves to other locations.

Single epidemiology studies cannot generally determine whether an observed effect is biologically related to the measured exposure unless the end point is unique and relatively rare, or the response is substantially elevated over background. Confidence in relating exposure with a health effect is increased if the effect is observed in multiple epidemiological studies supported by clinical (i.e., human) studies and laboratory animal studies. These latter studies, of course, must be conducted within certain ethical bounds.

The NAAQS assessment for humans initially focuses on the respiratory tract and uptake although the ultimate adverse effect may be at other sites. Air pollutants can have a variety of detrimental effects on the lung, including altered respiratory mechanics, reduced supply of oxygen, and increased stress, as well as other physiological effects such as a cardiovascular event, reduced resistance to infection, aging and chronic disease, and cancer. Because the possible health consequences span such a wide range, health researchers use a wide variety of measures to assess them. For example, mortality is typically reported as excess deaths, deaths per year, deaths per unit population, and similar measures. Morbidity may be detailed in studies from reported hospital admissions, reduced lung function, increased absences from school or work, and similar measures.

Studies of air pollutants also involve short-term and long-term exposures as well as exposure to high and low concentrations; exposure can also vary significantly with time. These exposures are primarily measured using ambient air monitoring equipment. Today, the EPA and the states operate a nationwide monitoring network that continuously tracks concentrations of several criteria pollutants, including PM, in the nation's ambient air. The network was established to allow the EPA and the states to determine compliance with the NAAQS. Ambient monitoring data can also be used to estimate average population exposures; however, this use is limited because of the population mobility and the fact that people spend large portions of their time each day indoors, where pollution concentrations may differ significantly from the outside air. In order to better estimate true exposures, researchers use techniques such as personal monitors and detailed activity pattern studies. Unfortunately, these are used less frequently in air pollution studies because the cost is high.

For the above reasons, a NAAQS can take various forms depending upon factors such as the nature of the health effect, exposure patterns, and the quality and quantity of the data used to determine compliance. A typical NAAQS may consist of a concentration level (usually expressed in parts per million or micrograms per cubic meter), an averaging time (e.g., a 1-hour, 24-hour, or annual average), a compliance statistic (e.g., the number of times a standard can be exceeded before it is a violation), and the length of the compliance period (e.g., a 3-year average).

The EPA promulgated the original NAAQS for PM in 1971, shortly after passage of the CAA and the establishment of the EPA. PM originally was defined as particles captured by a high-volume sampler, which collects particles up to about 45 microns. This fraction was designated total suspended particulate (TSP). In 1987 (52 FR 24854, July 1, 1987), the EPA changed the regulated pollutant to particles equal to or less than 10 microns in diameter. This fraction was referred to as PM_{10}. This change was made because it was learned that larger particles are not taken deeply into the lungs and, thus, are of less public health concern. As required by the CAA, the EPA continued to review and assess information necessary to determine whether further revisions to the PM NAAQS were required. However, when there was no further action by 1994, EPA was compelled to complete its review following a lawsuit filed by the American Lung Association (ALA). The EPA was ordered to complete its review and publish its findings on PM and ozone by early 1997. This due date was later changed to June 28, 1997.

On July 18, 1997, the EPA promulgated revisions to the PM NAAQS. The NAAQS for PM_{10} was retained with minor changes, but a new NAAQS was promulgated for particles equal to or less than 2.5 microns in diameter ($PM_{2.5}$). There are now two primary (i.e., health-based) standards for PM_{10}—an annual standard of 50 $\mu g/m^3$ and a 24-hour standard of 150 $\mu g/m^3$—and two primary standards for $PM_{2.5}$—an annual standard of 15 $\mu g/m^3$ and a 24-hour standard of 50 $\mu g/m^3$. The

$PM_{2.5}$ standard was based on the conclusion that smaller particles are taken even deeper into the lungs than PM_{10} and have a potential for more serious adverse health effects. This conclusion is largely supported by limited epidemiological studies that are the subject of considerable scientific controversy and that will be discussed in more detail below.

2. Indoor Particulate Matter

At the time of this writing, there was no federal legislation requiring the regulation of indoor air pollution with the exception of the workplace standards published by the OSHA. One difficulty in dealing with indoor air is that regulatory activities could potentially intrude on the individual's home and personal lifestyle which Congress and the federal agencies have been very reluctant to do. However, the OSHA did propose in April 1994 workplace standards on indoor air quality relating largely to environmental tobacco smoke. The proposal was based on the OSHA determination that employees working in indoor environments face a significant risk of material impairment to their health due to poor indoor air quality. The proposal was far-reaching and attracted over 100,000 comments and over 400 witnesses in public hearings. At the time of this writing in 1997, the OSHA continued to review the comments and testimony and no date was set for further action.

As noted above, the workplace regulatory development process used by the OSHA is similar to that used by the EPA, although adverse health effects in the workplace are often easier to link to specific substances. This is due to the fact that workplace exposure concentrations tend to be greater than outdoor exposure concentrations, and exposed populations and exposure times are much more consistent. Human health effects resulting from workplace exposure to air pollutants again rely heavily upon workplace epidemiological studies. While studies of small populations can often be used, there are still issues of genetic variability, health, and personal lifestyle. In fact, a drawback to many workplace studies is that they are often limited to a generally healthy, predominantly male workforce. This factor limits the ability of epidemiologists to extend the results to other populations which might include children, the aged, and the infirm. While a PEL assessment for humans focuses initially on the lung, other concerns may arise because of the generally higher concentrations of the substance exposures.

The original OSHA PELs included ceiling values and 8-hour time weighted averages (TWAs). The ceiling was a maximum concentration that was not to be exceeded at any time. The 8-hour TWAs factored in a worker's exposure across a typical 8-hour shift, 5-day week. Computation of the TWA exposed concentration is accomplished by multiplying exposure concentration by exposure time during each segment of an 8-hour work period and dividing the total by 8 hours. The 1989 revised PEL standard (since vacated) added a short-term exposure limit (STEL) which was defined as the employee's 15-minute TWA exposure which could not be exceeded at any time during a work day.

Measurement of workplace compared to outdoor air exposures is generally easier because the exposure concentrations are usually higher and more uniform. Today, there is a wide range of workplace air monitoring equipment, much of it portable and able to be attached to a worker's clothes to monitor actual exposure more closely. These devices provide useful data for establishing new standards and evaluating the effects of old standards.

Since 1971, the OSHA has maintained a list of 470 PELs for various forms of approximately 300 chemical substances, many of which are widely used in industrial settings. These PELs were based on research conducted primarily in the 1950s and 1960s and, for many of the substances, drew heavily on a similar listing established by the American Conference of Governmental Industrial Hygienists (ACGIH). The ACGIH is a professional society founded in 1938 with membership limited to professional personnel in governmental agencies or educational institutions engaged in occupational safety and health programs. While not governmental, the ACGIH's recommended guidelines were applied widely before the OSHA and still are used by many state agencies and others to protect workers.

Believing that the original PELs did not adequately protect worker health, the OSHA promulgated in 1989 (54 FR 2920, January 19, 1989) revisions to 212 existing exposure limits and limits for 164 new substances. In 1992, the OSHA further proposed to apply these standards to the construction, maritime, and agricultural sectors. These actions resulted in a lawsuit and, in 1992, the 11th Circuit Court of Appeals vacated the standards (*AFL-CIO v. Secretary of Labor*, 965 F.2d 962 [11th Cir. 1992]) and ruled that the OSHA did not sufficiently demonstrate that the new PELs were necessary or that they were feasible. This decision forced the OSHA to return to its original 1971 limits. The OSHA has currently assigned a high priority to the revision of out-of-date PELS.

The regulation by the OSHA of PM in the workplace currently includes PELs for specific chemical substances that may exist as particles in the workplace air. Examples are certain elements and their compounds, metal dusts, carbon black, cotton dust, silica dusts, silicates, a miscellaneous category called inert or nuisance dust, and asbestos.

III. RISK ASSESSMENT OF PARTICULATE MATTER

A. Introduction

As summarized in Chapter 2 and described in detail in Chapters 3 through 6, risk assessment consists of four steps: hazard identification, dose–response assessment, exposure assessment, and risk characterization. The hazard identification step determines whether a substance is related to an adverse health effect. The dose–response assessment step determines the relation between the magnitude of the exposure and the likelihood of occurrence of the health effect in question. The exposure assessment determines the extent of human exposure both before and after controls. Finally, the risk characterization step combines all of the preceding information and describes the nature and the magnitude of the human risk, along with all applicable uncertainties.

B. Characteristics that Influence the Particulate Matter Risk Assessment

As indicated by the PM NAAQS, the health effects of PM are believed to be strongly related to the size of the particles inhaled, because the size and composition determine behavior in the respiratory system (e.g., how far the particles penetrate, where they deposit, and the effectiveness of the body's clearance mechanisms among other factors). Particle size is also an important factor in determining atmospheric lifetime. Based on observed particle size and formation mechanisms, PM is usually classified into two fundamental modes: fine and coarse particles, with the cut point between the two at about 1 to 3 microns (as noted above, the EPA chose 2.5 microns). Importantly, fine and coarse particles appear to be differentiated by their sources and formation processes, chemical composition, solubility, acidity, atmospheric lifetime and behavior, and transport distances. For example, fine particles are generally formed from gases while coarse particles are generally directly emitted as particles. In addition, fine particles have a longer atmospheric lifetime than coarse particles. One result is that exposure to PM indoors in the U.S. is often to smaller particles that are generally more concentrated—and whose concentrations are more consistent—than outdoor exposure. Another important factor is that since the oil crisis of the early 1970s, homes and other buildings have been modified or built to reduce energy costs through minimization of air movement between the indoors and outdoors. Effectively sealing rooms reduces the infiltration of outdoor PM, but can correspondingly result in increased indoor concentrations because there is less exfiltration.

The original development of the PM NAAQS depended, and its ultimate implementation depends, in large part on the atmospheric concentrations of PM measured by a nationwide network of atmospheric monitors operated by the EPA and state and local air pollution agencies. Extensive data on PM_{10} have been available since mid-1987 when the PM_{10} NAAQS was first promulgated. However, data on $PM_{2.5}$ was limited at the time that the PM NAAQS was promulgated, and $PM_{2.5}$ concentrations often had to be estimated from other data, including PM_{10} concentrations and visibility data. The distribution and composition of PM vary widely by location in the country, being influenced by man-made sources, natural sources, and weather; these variations can significantly affect the risk assessment.

C. Hazard Identification

1. Evidence of Mortality Associated with Exposure to Particulate Matter

The earliest substantiated reports of excess mortality from short-term exposures to community air pollution containing high levels of PM come from several air pollution disasters, including the Meuse Valley in Belgium (1931), Donora, Pennsylvania (1949), and London, England (1954). In these disasters, winter weather inversions led to very high (e.g., 500–1,000 $\mu g/m^3$ in London) PM and SO_2

concentrations which were associated with large simultaneous increases in morbidity (i.e., illness) and mortality (i.e., death). In one follow-up study, survivors with either chronic disease prior to the episode or who became acutely ill during the episode were found to have higher subsequent rates of mortality and morbidity. Later studies in London also showed a continuum of response across a full range of PM levels, suggesting effects at levels commonly observed in the U.S. ambient air. However, these data must be interpreted cautiously. For example, the analyses considered only exposures to PM and SO_2. Yet the air pollution resulted predominantly from coal combustion and, thus, the population was also exposed to emissions of nitrogen oxides (NO_x), carbon monoxide (CO), and other potentially toxic emissions, which were not accounted for. In addition, studies have shown that average Americans spend as much as 90% of their time indoors even in good weather. During times of air pollution emergencies, it may be logical to assume that people will spend even more time indoors. We also know that most of the mortality and illness occurs indoors. Thus, the analyses are comparing measured outdoor concentrations of two specific pollutants against mortality and illness perhaps more associated with indoor exposures to a wide range of substances at varying and generally unknown concentrations.

In the 1980s, as a result of the growing availability of PM_{10} monitoring data and newer statistical techniques, a number of short-term studies of mortality and illness and longer-term studies of mortality associated with PM exposures were published. Importantly, these studies reported statistically significant positive associations between short-term exposures to PM (measured as TSP and PM_{10}, and a limited amount of $PM_{2.5}$) and mortality. As reported in the EPA Criteria Document, of 38 studies published between 1988 and 1996 "most found statistically significant associations between increases in ambient PM concentration and excess mortality . . . [even though] these locations differ significantly in pollution and weather patterns." However, these studies cannot determine with certainty whether an individual component of ambient air exposure caused the increased mortality or whether it was the complex of air pollutants as a whole.

Prior to 1990, cross-sectional studies were generally used to evaluate the relationship between mortality and long-term exposure to PM. In some cases, these studies showed statistically significant positive associations between higher long-term PM concentrations and higher daily mortality rates across communities. However, these studies did not typically account for other important risk factors that could be associated with an increased risk of mortality, including smoking, lifestyle, and exposure patterns; they accounted for the effects of weather and other air pollutant variables only in a limited way, which limited their usefulness. Since 1990, more studies have taken into account these other risk factors. In these studies, groups of individuals are chosen and detailed information on a number of variables likely to be important to the assessment is gathered. Unfortunately, while these studies significantly improved the ability of the study to isolate the effects of exposure to air pollution, the detailed information on other lifestyle risk factors was rarely complete and generally focused only on a few obvious factors, such as smoking, age, sex, and race. In addition, detailed information on exposure to air pollution is generally available only from centrally located air pollution monitors in the geographic areas from which the groups of individuals are drawn. Nonetheless, these studies are reported by the EPA as contributing evidence to the hypothesis that increased exposure to PM is associated with increased mortality.

The EPA used several short-term studies and two long-term studies, described in previous chapters, to support the conclusions that led to the revised NAAQS. The most extensive long-term study is referred to as the Harvard Six Cities Study (Dockery et al. 1993); another important study is referred to as the American Cancer Society (ACS) Study (Pope et al. 1995). These studies utilized personal data on individuals for variables such as smoking, and regional ambient monitors for data on exposure. One difference between the short-term studies and long-term studies is that inferences in short-term studies are generally based on differences in ambient levels of pollutants from day to day while inferences in the long-term studies are generally based on differences in levels of pollutants from city to city. Unfortunately, a number of other easily accessed risk factors, such as weather and exposure to other pollutants, were not controlled for in either the Harvard Six Cities or ACS studies.

In the Harvard Six Cities study, several thousand people were followed for 14 years in six cities. Information was gathered regarding smoking, education level, occupation, and other potentially important risk factors. After adjustment, elevations were reported in several measures of long-term PM concentration that were significantly associated with total mortality rates. The adjusted increase in risk of 26% (95% confidence interval of 8–47%) from PM exposure was nearly equal for PM_{10}, $PM_{2.5}$ (although $PM_{2.5}$ exposure data were limited), and sulfates between the cities with highest and lowest air pollution.

In the ACS study, over one-half million adults in 151 U.S. cities were studied in an attempt to test the relationship between long-term exposure to fine particles and increased mortality. This study was designed to follow up on a suggestion from the Harvard Six Cities study that long-term exposure to fine particles is associated with increased mortality. To test the hypothesis, the association between multiyear concentrations of two fine particle indicators, $PM_{2.5}$ and sulfates, was evaluated. As in the Harvard Six Cities study, information for each individual was used to adjust for other important risk factors, including age, sex, smoking, passive smoking, and occupation. After adjustment for the other factors, $PM_{2.5}$ concentrations ($PM_{2.5}$ data were limited and concentrations generally were estimated by adjusting TSP or PM_{10} data) were reported to be associated with a 17% (95% confidence interval of 9–26%) increase in total death rates, and sulfate concentrations were associated with a 15% (confidence interval 5–26%) increase in total death rates. Finally, the ACS study showed somewhat lower relative risks of mortality than the Harvard Six Cities study between the most-polluted and least-polluted cities for the total population and selected smoking groups. In summary, the two key studies demonstrate small observed increases in mortality with increased PM exposure but relatively large error bands.

There are a number of uncertainties associated with these studies. First, the lack of consideration of exposures to other criteria air pollutants for which data were available is a serious flaw because air pollution is a complex mixture of substances, exposure to each of which or the complex mixture of which may present an independent risk factor. In most instances, data on other pollutants were available, but were not gathered and evaluated. Weather and seasonal differences also were given limited consideration. For example, the effects of air pollution can be modified by season because the mix of pollutants and patterns of outdoor activities vary. In a given season, mortality can also be affected by weather conditions such as temperature, both hot and cold. At least three independent investigators examined five of the cities' data with control of other pollutant variables and seasonal changes. When other air pollutant variables are considered impartially, the PM association with daily mortality and morbidity does not stand out. Different pollutants also demonstrate different associations for various seasons in different cities, and their analyses indicate that different pollutants are related to mortality and morbidity in different seasons and that, again, PM does not stand out. The EPA also gave little attention to the potential adverse effects from exposure to CO and nitrogen dioxide (NO_2). Another study found significant associations between CO exposure and hospital admissions for congestive heart failure. In addition, a paper published in *Epidemiology* found significant associations between NO_2 and daily mortality rates in Philadelphia and with hospital admissions for respiratory diseases in Minneapolis. One researcher also showed that other reasons could account for the correlation, including the fact that the proportion of the population with a sedentary life-style correlates well with the adjusted mortality rates in the Six Cities study. There also are significant questions concerning the actual exposures of the populations. The studies generally assumed that exposure was determined by the PM monitors located in the urban areas that were studied. Yet, the people in those urban areas were, in fact, likely to be exposed quite differently because they spent considerable time indoors and in transit. In addition, most of the mortalities were older or ill persons who were confined to homes or hospitals, many with filtered air systems. Finally, at the time this book was written, several researchers believed that the adverse effects of PM are more likely due to short-term peaks rather than long-term averages. In an editorial letter in *Science* (December 5, 1997), one researcher commented that "attributing PM effects to a 24-hour average . . . is like attributing daily mortality in a war zone to the 24-hour lead concentration instead of bullets."

2. Evidence of Life Span Shortening

In preparing the Criteria Document, the EPA attempted to evaluate the potential shortening of life span associated with PM exposure in these and other studies. The document states that epidemiological studies "suggest [that] ambient PM exposure affects mortality both in the short and long term, and promotes potentially life-shortening chronic morbidity in the long term." However, the EPA concluded that it was not possible to confidently estimate the number of years of life lost.

3. Evidence of Increased Illness (Morbidity)

In addition to the effects on mortality rates, the Criteria Document assessed the potential for increased morbidity with increased exposure to PM. In a review of 13 epidemiological studies, 12 were reported as showing statistically significant positive associations between short-term exposures to PM and hospital admissions for res-piratory-related and cardiac diseases. As with the mortality studies, these results were observed in communities across the U.S. However, although studies were reported as showing consistent statistically significant associations between such measures of morbidity and increased short-term levels of indicators of PM, the EPA admitted that the studies were difficult to interpret.

4. Evidence of Decreased Lung Function

The EPA reported that community epidemiological studies showed that PM exposure is associated with altered lung function and increased respiratory symp-toms. Effects on respiratory mechanics range from mild transient changes with little direct health consequence to incapacitating impairment of breathing. For example, asthmatic subjects appear to be more sensitive than healthy subjects to the effects of acid aerosols on lung function, and fine aerosols may alter lung function to a greater degree than larger aerosols. However, laboratory studies of animals using acid aerosol exposures at concentrations up to 1,000 $\mu g/m^3$ did not produce direct changes in pulmonary function in healthy animals except in guinea pigs. The EPA did conclude that other studies reported positive associations between respiratory symptoms and PM pollution that lead to concerns about the longer-term potential for increases in the development of chronic lung disease.

5. Evidence of Sensitive Population Groups

The Criteria Document indicates that several subgroups of the population are "apparently more sensitive [susceptible] to the effects of community air pollution containing PM." These groups include individuals with respiratory and cardiovas-cular disease, individuals with respiratory infections, children (likely due to both greater exposure and higher ventilation rates), and asthmatics. Various studies have examined the elderly, but those results are reported inconclusive by the EPA.

6. Evidence from Animal and Occupational Studies

Animal and occupational studies were used to investigate the likelihood for alteration of lung tissues and components as a result of PM and acid aerosol exposures. Such changes were noted in some studies. For example, alterations were clearly noted in extensive studies of single and multiple exposures to sulfuric acid aerosols, but only at high concentrations (i.e., greater than 1,000 $\mu g/m^3$). Other studies summarized in the Criteria Document, including analyses of silica and natural dust exposures, show fairly specific lung effects at relatively high concentrations, but inconclusive effects on body defense mechanisms.

7. Evidence for Mechanisms of Effect

The Criteria Document postulates several physiological and pathological mech-anisms for responses to exposure to PM. However, these mechanisms are largely derived from animal studies conducted at exposure levels generally greater than those found in the ambient air; the Criteria Document indicates that these studies generally were not well controlled. Thus, the EPA concluded that "at present, available toxicological and clinical information yields no demonstrated biological mechanism(s) that can explain the associations between ambient PM exposure and mortality and morbidity reported in community epidemiology studies."

8. Scientific Review of the Health Hazards

The CASAC completed its review of the EPA Staff Paper on June 13, 1996. The CASAC concluded in its closure letter to the EPA Administrator that "although our understanding of the health effects of PM is far from complete, the Staff Paper, when revised, will provide an adequate summary of our present understanding of the scientific basis for making regulatory decisions concerning PM standards." Of the 21 Panel members, 17 voted to approve the report, 2 voted against approval, and there were one abstention and one absence. However, the panel requested that the EPA make significant changes in the final document. Those changes were articulated to the EPA staff at the meeting, and in writing, and appear to have been appropriately addressed by the EPA prior to proposal of the revised PM NAAQS. A majority of the CASAC recommended keeping the present 24-hour PM_{10} NAAQS at the current level with a possible change in form, and there was a consensus that a separate new $PM_{2.5}$ NAAQS should be established at a 24-hour and/or annual standard, although there was no consensus on the level, averaging time, or form of the standard. The final CASAC letter to the EPA Administrator provided the following rationale for a new $PM_{2.5}$ NAAQS:

"[There is] strong consistency and coherence of information indicating that high concentrations of urban air pollution adversely affect human health, there are already NAAQS that deal with all the major components of that pollution except $PM_{2.5}$, and there are strong reasons to believe that $PM_{2.5}$ is at least as important as $PM_{2.5-10}$ in producing adverse health effects."

In later testimony before a Senate Committee, the Chairman of the CASAC and a former chair (and consultant during the PM review) highlighted the division in the CASAC on the EPA proposed $PM_{2.5}$ standards. The Chairman said that he "could not endorse them." The consultant said during his testimony that the proposal was a "prudent step to protect public health." Both scientists, however, agreed that the CASAC did reach consensus on the need for some unspecified form of a $PM_{2.5}$ standard.

In conclusion, while epidemiology studies appear to show a relationship between PM exposures and excess death and illness, the studies are generally flawed by the lack of consideration of important risk factors, and they provide insufficient information to establish a cause for the adverse effects that are identified. Compounding this is the lack of any biologically plausible mechanism to explain the supposed PM health effect.

D. Dose–Response Assessment

As noted earlier, the dose–response step seeks to determine the relation between the magnitude of the exposure and the likelihood of occurrence of the health effect in question. The analyses discussed in this chapter and in earlier chapters show that, to the extent that toxicity is associated with PM, it is related to a number of variables, including particle size, composition, particle source, other exposures, and other biological factors (e.g., age and presence of preexisting disease). In other words, PM encompasses a wide range of substances that are physically similar and able to be inhaled. However, beyond that the adverse health effects that might result from exposure to the inhaled PM may be significantly different depending upon the precise size, composition, and source of the PM, as well as other factors. As might be expected from this and the many basic uncertainties described in the hazard identification chapter, the EPA Criteria Document notes that the characterization of the dose–response continuum from PM exposure is far from complete and concludes that the linkage between exposure, dose, and response remains weak and qualitative at best.

While there was insufficient data at the time this book was written to develop a dose–response curve for PM exposure from clinical, animal, and other toxicological test results, dose–response curves had been developed based on the results of the epidemiological studies. However, the use of these data is hampered because of uncertainties in variations in the extent of exposure of the population, the relative risk of mortality that the exposure confers, and the shape of the underlying exposure–response relationship. Available monitoring information provides rough estimates of exposures to PM_{10}, but data are much less extensive for $PM_{2.5}$. Relative risk ratios for short-term mortality studies are reported by the EPA as generally showing a 2 to 10% increase in risk of mortality over background risk, but the EPA admits that the data vary from site to site. Furthermore, the relative proportions of total PM mortality attributable to short-term and long-term exposure are not known.

In the face of this uncertainty, the EPA assumed that the relative risk of mortality increases linearly with the concentration of PM with no evidence of a threshold. This approach is generally the most health protective assumption and has been used broadly in the regulation of carcinogens which cannot be shown to exhibit a threshold mechanism. The no-threshold conclusion, however, is much more in question for PM. A major problem is that the dose–response relationship has been investigated largely through analysis of epidemiological studies of exposures in a relatively narrow range of exposure concentration. Furthermore, the use of laboratory animal toxicological data has been limited for many reasons, including the difficulties in extrapolating from animals to humans. Another problem is that people are almost always exposed to PM in conjunction with other pollutants and the possible effects of the combined exposures are rarely known. Finally, there have been only a few attempts to investigate possible exposure–response relationships other than a linear one. Yet, one study reported nonlinear associations between two pollutants, TSP and SO_2, and daily mortality in Philadelphia. In addition, the study reported that TSP had little effect on mortality below 100 $\mu g/m^3$.

In conclusion, with the absence of a dose–response curve from clinical, animal, or other toxicological test results, EPA chose the default relationship which assumes a linear response. As noted earlier in this book, the linear model is generally the most conservative (i.e., health protective), meaning that risk is not underestimated when using this model.

E. Exposure Assessment

Outdoor exposure to PM is systematically measured using a nationwide network of air pollution monitors operated by the EPA and state and local air pollution control agencies. Most cities have one or more PM air monitors; larger cities typically have more than one. The primary purpose of this network is to provide the information necessary to determine what areas of the country are meeting the PM NAAQS (called "attainment") and which areas are not (called "nonattainment"). Since 1987, a large number of PM_{10} monitors have been in place, and a growing number of $PM_{2.5}$ monitors are being installed to provide the data necessary to determine attainment with the PM NAAQS promulgated in 1997. While the total number of monitors is impressive, they do not necessarily provide an accurate measure of specific population exposures because the monitors are usually located at fixed sites in city centers, while people move more widely, travel frequently, and spend a majority of their time indoors.

In addition, the accurate measurement of ambient PM is, as noted by the EPA, challenging and expensive. It was noted earlier that $PM_{2.5}$ data were scarce at the time the EPA established the $PM_{2.5}$ NAAQS in 1997. Furthermore, few specific PM chemical constituents beyond sulfates, nitrates, and a few organics have been widely monitored or epidemiologically assessed. On the positive side, because of their longer atmospheric lifetime, $PM_{2.5}$ particles appear to be more uniformly distributed than coarse particles within a specific urban airshed. In addition, particles can infiltrate indoors, but fine particles are removed less rapidly indoors than coarse particles, leading to higher and more consistent concentrations.

Models have also been used to estimate outdoor exposures to PM; however, most are subject to uncertainty. Typical problems include seasonal adjustments, adjustments for copollutants (e.g., SO_2, ozone, and CO appear to play a role in modifying PM's adverse effects), and adjustments for weather (e.g., at this time few studies have examined possible statistical interactions between weather and air pollution).

Indoor exposure to PM has not been systematically measured, although considerable study has focused on respirable suspended particles indoors, particularly environmental tobacco smoke (ETS). Indoor PM exists in both solid and liquid phases and can arise from many sources, including mold spores, pollen, human and animal dander, dusts, combustion exhaust, inorganic aerosols, consumer products, and others. ETS is an important contributor and likely exists as liquid or waxy droplets that may also contain some amount of ash. With time, the more volatile components of the smoke evaporate and the particles become smaller and comprised of higher molecular weight materials.

As discussed in Chapter 8, four large-scale studies have been performed that included investigation of indoor and outdoor PM exposures: (1) the Six Cities Study, discussed above in the section III.C.1, took PM measurements in over 1,400 homes over about ten years; (2) the New York State Energy Research and Development Authority (ERDA) carried out a study in 433 homes in two New York counties; (3) the EPA Particle TEAM study investigated 178 homes in California; and (4) additional studies of ETS were conducted in a number of European cities. These studies have provided a large data base of information on exposure to PM over a range of conditions, seasons, ages, and other variables. In general, cigarette smoking is the largest single contributor to respirable PM exposure, and indoor PM concentrations are typically higher, often double, those of concurrent outdoor concentrations. Furthermore, one study showed that concentrations measured by personal monitors (i.e., worn on the person near the breathing zone) showed even higher concentrations than indoor concentrations measured by fixed monitors.

Exposure is a function of a number of factors, including breathing rate, particle size, composition, and concentration. Ventilation rates range widely, but are typically higher in children and active adults, and lower for all ages at night. As noted earlier, smaller particles are inhaled more deeply, and certain composition-related factors lead to increased dose. Indoor and outdoor concentrations of PM were measured in the Harvard Six Cities Study. The researchers found that indoor concentrations were higher than outdoor concentrations, except in one city, and noted that a major source of indoor PM was cigarette smoke. Respirable PM concentrations ranged from lows of 10–20 $\mu g/m^3$ indoors and outdoors to highs of over 300 $\mu g/m^3$ indoors and about 60 $\mu g/m^3$ outdoors. The New York ERDA study focused on the effects of home heating systems and measured indoor $PM_{2.5}$ concentrations in the range of 25–35 $\mu g/m^3$ and outdoor concentrations in the range of 15 $\mu g/m^3$ (in largely suburban areas). In one series of EPA TEAM studies, indoor $PM_{2.5}$ concentrations ranged from about 50 to about 200 $\mu g/m^3$; outdoor PM_{10} concentrations were somewhat higher, ranging from about 60 to about 220 $\mu g/m^3$; and personal monitoring concentrations were higher still, ranging from 70 to about 270 $\mu g/m^3$. In the warmer California climate, the TEAM study showed a much higher contribution of outdoor PM to the total indoor concentration than the Harvard six cities, which were located in Wisconsin, Ohio, Massachusetts, Tennessee, Missouri, and Kansas.

In conclusion, exposures to indoor air pollutants are frequently higher indoors than outdoors, particularly in homes and buildings with sources of, or mechanisms of entry for, PM.

F. Risk Characterization

The EPA concluded that there is substantial evidence that ambient PM, alone or in combination with other commonly occurring pollutant gases, is associated with "small but significant increases in mortality and morbidity in sensitive populations at concentrations below the levels of the current ambient standards for PM." This conclusion was largely supported by the Harvard Six Cities and ACS epidemiology studies. The EPA also believes that the body of evidence suggests a biological link between PM and increased mortality rates, but admits that supporting evidence for plausible mechanisms of action is lacking in the published literature. The EPA also believes that the evidence is considerably stronger for fine particles (2.5 microns and less) and, while there is ample reason to be concerned with coarse particles (2.5 to 10 microns), there is less direct evidence regarding the potential effects of coarse particles. The EPA, therefore, concludes that coarse particles are either less potent or a poorer surrogate for community effects than fine particles. The EPA utilized a risk assessment approach based on the ACS Study to establish a dose–response relationship for fine particles. The selected relationship was a linear dose–response curve. Using this approach, the EPA estimated that the full attainment of the $PM_{2.5}$ standard would result nationwide in the yearly avoidance of 1,000 to 6,000 incidences of premature death and 22,000 new cases of chronic bronchitis. However, given the many uncertainties described here both in the hazard assessment and the dose–response assessment, the real increase in mortality and morbidity resulting from long-term exposure to PM either indoors or outdoors could be much lower than EPA's estimate, and possibly there could be no adverse effect.

G. Summary

Although many NAAQS revisions are surrounded by controversy, the EPA PM revisions led to some of the sharpest disagreements ever. In part, the revisions were driven by external forces because the ALA filed suit in 1994 to compel the EPA to complete the review of the PM NAAQS by December 1995. Although the EPA argued that a decision should not be required until the science was clear, the Court said the mandate of the CAA was clear and ordered the EPA to complete its review. A schedule was specified with a final published decision required by June 28, 1997. The issue was also widely debated with frequent articles in leading health journals such as *Epidemiology*, the *New England Journal of Medicine*, and *Inhalation Toxicology*.

The EPA revisions to the PM NAAQS were based largely on epidemiological studies that appear to show a relationship between PM exposures and excess mortality and morbidity, and a growing belief by the EPA that a further division of the PM_{10} NAAQS is necessary to protect the public with an adequate margin of safety. However, as described above, the epidemiology studies appear flawed by the lack of consideration of other important risk factors and many believe that they provide insufficient evidence to establish a cause for the adverse effects that are identified. The most important shortcomings are the following:

- Exposure to other significant air pollutants (for example, CO and NO_x) generally was not considered in the epidemiological studies.
- The studies assume that ambient air monitoring data from a limited number of community monitoring stations adequately describe total personal exposure. There is also a lack of measured $PM_{2.5}$ data and the use instead of TSP and PM_{10} data adjusted using invalidated conversion factors.
- At least three independent investigators reanalyzed data from five of the cities and came to different conclusions. They found that when other pollutants were controlled for in the analysis, no single pollutant emerged as responsible for the health effect. They also showed that the effects differed by season of the year.

- The raw data from the Harvard Six Cities and ACS studies, funded in part by the EPA, had not been released by the researchers at Harvard University who conducted the analyses at the time of this writing. These were the only studies performed with $PM_{2.5}$ data and, without the raw data, the studies could not be meaningfully evaluated by other researchers.

Adding to the debate is the current lack of any biologically plausible mechanism to explain the supposed PM–mortality relationship. In addition, the statistical differences reported are small. For example, in the Harvard Six Cities study small differences in the ages of the groups studied could have accounted for the differences attributed to PM, but were apparently not considered. There continues to be considerable debate about the EPA establishment of the 2.5 micron cutoff. Some scientists believe that a 10 micron cutoff is sufficient, others believe that a 2.5 micron cutoff is more appropriate, and still others believe that a smaller cutoff for particles less than 1 micron is appropriate. Importantly, the EPA promulgated the new $PM_{2.5}$ standard with very little nationwide data on $PM_{2.5}$ and its possible association with excess mortality and morbidity. Table 9.1 provides a list of the information important to the PM NAAQS decision and an estimate of the current scientific confidence in the accuracy of that information.

Table 9.1 Estimate of the Current Scientific Confidence in Information Important to the Particulate Matter NAAQS Decision

Information Important to NAAQS Decision	Current Scientific Confidence in the Information
Level of the PM_{10} standard	Moderate
Effects of long-term exposures at high concentrations in humans	Moderate
Level of the $PM_{2.5}$ standard	Moderate to low
Selection of 2.5 micron cutoff	Moderate to low
Exposure/risk analysis	Moderate to low
Epidemiology results	Low
Biologically plausible mechanism for the apparent PM effect on humans	Low

BIBLIOGRAPHY

Dockery, D., Pope, C., et al. 1993. An association between air pollution and mortality in six U.S. cities, *New England Journal of Medicine* 329:1753–1759.

Environmental Protection Agency (EPA). 1996a. *Air Quality Criteria for Particulate Matter, Volumes I–III*, EPA/600/P-95-001aF–EPA/600/P-95-001cF, April 1996.

Environmental Protection Agency (EPA). 1996b. *Review of the National Ambient Air Quality Standards for Particulate Matter: Policy Assessment of Scientific and Technical Information*, EPA/452/R-96-013 (OAQPS Staff Paper), July 1996.

Pope, C., Thun, M., et al. 1995. Particulate air pollution as a predictor of mortality in a prospective study of U.S. adults, *Am. J. Respir. Crit. Care Med.* 151:669–674.

CHAPTER 10

Future Directions in Risk Assessment

David R. Patrick

CONTENTS

I. INTRODUCTION AND PURPOSE

The understanding and use of environmental risk assessment has grown rapidly since the National Research Council (NRC) established its guiding principles (NRC 1983). As noted in Chapter 2, the NRC first identified and described the four steps of environmental risk assessment, namely, hazard identification, dose–response assessment, exposure assessment, and risk characterization. The use of each of these components in indoor air risk assessments is discussed fully in Chapters 3 through 6. These chapters identify numerous areas of uncertainty in conducting risk assessments, including variations in the models used, variations in the inputs to the models, inexact knowledge of the underlying science, and natural variability. Chapter 7 considers more broadly the uncertainties of risk assessment, and Chapter 8 describes measurement methods and results for indoor air pollutants. Substantial research is either underway today or planned for the future that will address many of these subjects. The purpose of this chapter is to describe some of that research and project how its successful conclusion might alter the dimension and use of risk assessment to understand and benefit indoor air quality.

II. INDOOR AIR RISK ASSESSMENT RESEARCH PROGRAMS

A. U.S. Environmental Protection Agency

The EPA research into indoor air pollution began in the late 1970s. For example, the original TEAM (Total Exposure Assessment Method) studies sought to understand better the distinctions between outdoor and indoor air, and they found that indoor exposures to many air pollutants were significantly greater than expected. As described in Chapter 8, the TEAM studies continued for many years.

In its 1989 Report to Congress on Indoor Air (EPA 1989), the EPA described the ambitious indoor air research program required by Title IV (Radon Gas and Indoor Air Quality Research Act) of the 1986 Superfund Amendments and Reauthorization Act (SARA). Title IV provided for the first time a Congressional mandate for a national indoor air research program. SARA Title IV specifically required research into identification, characterization, and measurement of sources and levels of indoor air pollution; development of instruments for indoor air quality data collection; the study of high-risk buildings; identification of the effects of indoor air pollution on human health; development of mitigation measures to prevent or abate indoor air pollution; demonstration of methods for reducing or eliminating indoor air pollution; development of methods for assessing the potential for radon contamination of new construction; and examination of design measures to avoid indoor air pollution. However, during the years from enactment of SARA Title IV to the time this book was written in 1997, no legislative program was enacted to regulate indoor air quality and the EPA budget allowed for only portions of the mandated research program. For example, the EPA focused in the early 1990s on developing information useful for reducing exposure to unhealthy levels of indoor air pollutants; this effort used voluntary approaches and partnerships to educate people from building managers to consumers to the problems of indoor air quality and appropriate solutions. The research focus at that time was development of information to be used in preparing guidance about reducing the health risks of indoor contaminants, including radon, second-hand tobacco smoke, and emissions from building and consumer products.

In order to meet the mandate of SARA Title IV, the EPA (EPA 1989) identified several "need" categories, including the following that are directly related to indoor air risk assessment:

- Risk assessment methodology needs, which focus on health and hazard identification, dose–response assessment, exposure assessment, and risk characterization frameworks and methods, especially as they relate to the comparability of results from oral vs. respiratory toxicity studies.

- Exposure assessment and modeling needs, including methods development and evaluation, measurement studies, development of predictive models, and the management of measurement data.

Much of the EPA planned indoor air pollution research, including the work on risk assessment methodologies and exposure assessment and modeling, was to be coordinated with other organizations such as the Department of Health and Human Services (DHHS), the Department of Energy (DOE), the National Institute for Science and Technology (NIST), and the Consumer Product Safety Commission (CPSC) in the Federal government, along with many states and the private sector.

B. Center for Indoor Air Research (CIAR)

The CIAR is a nonprofit corporation formed in the U.S. in March 1988 to sponsor research on indoor air issues and to facilitate communication of research findings to the scientific community. The Center utilizes a Science Advisory Board, consisting of experts in health, science, and architecture, to develop its research agenda and to recommend proposals for funding. The proposals are submitted by qualified individuals or organizations and evaluated by a large number of scientific and technical peer reviewers prior to submittal to the Science Advisory Board. This process seeks to ensure that research is funded which can contribute to the knowledge bank on indoor air.

In a 1996 publication (CIAR 1996), CIAR described its 1996–1997 research agenda. Research needs were grouped according to sources investigated, exposure/dose assessment, health effects, perception of indoor air quality, and engineering control strategies. Contaminants of interest included volatile organic compounds (VOCs), environmental tobacco smoke (ETS), biological aerosols (e.g., aeroallergens and aeropathogens), and particulate matter. The publication stated CIAR's interest in all relevant chemistry, physics, control strategies for, health effects caused or aggravated by, and psychosocial factors influencing, the perception of indoor air quality.

Sources needing CIAR research include cooking, consumer products including pesticides, heating and cooling systems, building materials, and electronic equipment. In addition, the distributions of sources and chemicals can be important. For example, some toxicologically significant compounds are being studied within risk assessment frameworks, but much work remains in characterizing the distributions of various agents in specific environments and assessing their impacts on human health. Research on biological agents was identified as a specific area of need.

Exposure assessment and dosimetry are key CIAR research areas that assist in determining the health consequences of exposure. In particular, understanding the effect of aerodynamic respiratory tract defenses and complex particle-gas compositions is important.

Health effects and responses also are important CIAR research areas. One major ongoing question is the validity of point or time-weighted measures. In other words, for a given, well-characterized indoor environment, do measurable health effects relate to cumulative, chronic, low-level concentrations, acute peak concentrations, and/or synergistic effects between substances? Better elucidation of health responses to interactive, low-level, complex exposures is needed, along with better definition of specific health responses resulting from specific exposures.

Perception of indoor air quality also continues to be a CIAR research need. While there has been substantial progress in developing techniques to measure contaminant concentrations, more research is needed to quantify human responses to indoor air environments. Studies have shown that worker health can be influenced by individual, perceptual, psychosocial, and psychophysical factors.

Finally, CIAR research is planned on engineering controls of indoor air quality to help reduce adverse health effects. The choice of an engineering control strategy depends on psychosocial and psychophysical influences as well as upon measurable contaminant concentrations. Thus, control also necessitates developing a knowledge of "healthy building characteristics."

C. Other Organizations

A number of other organizations conduct indoor air pollution research, specifically on methods related to risk assessment. Several federal agencies have indoor air responsibilities and conduct research, including the Bonneville Power Administration, the CPSC, the DOE (e.g., the Office of Conservation and Renewable Energy), the DHHS (e.g., the Office on Smoking and Health), the Tennessee Valley Authority, and the DOE national laboratories (e.g., Lawrence Berkeley Laboratories and Oak Ridge National Laboratory).

Some private and professional organizations also conduct research and/or develop management and control guidelines. The American Conference of Governmental Industrial Hygienists (ACGIH) is one of the best known professional organizations which develops and revises workplace exposure guidelines. Others like the American Industrial Hygiene Association (AIHA) and the American Society of Heating, Refrigerating, and Air-Conditioning Engineers (ASHRAE) play leading roles in establishing methods and guidelines. Product manufacturers also conduct research to identify new products and materials that are associated with fewer indoor air quality problems. Many of the manufacturers are represented by trade associations that also fund research activities.

Finally, many colleges and universities in the U.S. conduct important research on indoor air quality, and considerable research is being conducted in Canada, Europe, and other countries. The references provided in the various chapters in this book provide a number of examples of important research being conducted here and abroad.

III. INDOOR AIR RISK ASSESSMENT RESEARCH NEEDS

A. General Risk Assessment Research

The understanding and use of risk assessment continues to grow since its inception in the 1970s and it continues to be the subject of considerable research. Much of that research is focused on reducing the many uncertainties and in gathering data to allow more pollutant, source, and site-specific data to be used rather than traditional, typically conservative, default values. Conservative default values are often necessary initially when there is incomplete knowledge of the mechanisms of toxicity or other factors. Unfortunately, large gaps remain in the scientific knowledge in many areas; thus, the use of conservative default values continues. For example, substantial research on hazard identification is being undertaken to improve the categorization of cancer weight-of-evidence from animal and short-term studies, to determine whether or not cancers of varying types or severities should be given equal weight, and to develop a better understanding of the mechanisms of carcinogenesis and other toxic effects.

Research on dose–response assessment is underway to identify more appropriate models than the linear nonthreshold model initially used as the default because it provided the most conservative, plausible estimate of risk. Biologically based models are being developed that provide more accurate risk estimates. Other research focuses on determining the effective dose at the site of injury, as opposed to the exposed or intake dose. This also requires a better understanding of the biological processes and the distribution of chemicals through the body. Pharmacokinetic approaches utilize mathematical modeling to predict these processes, but to date have been developed for only a few specific chemicals. Where developed, the modeling shows a decrease in expected risks because smaller quantities of the chemicals are typically delivered to the target organs than are present at the point of human contact with the chemical. Another area of interest involves episodic exposures. In these cases, the mathematical models can predict the half-life of chemicals in the body that are metabolized or excreted.

Research on exposure assessment includes the development of more site-specific information to replace the highly conservative assessments of the early days; these often estimated maximum individual risks based on the assumption that a person could be exposed continuously (i.e., 24 hours per day, 365 days per year, 70 years) to the worst-case ambient concentration resulting from the emission of a pollutant of concern. Although generally unrealistic, governmental regulators at the time had no better basis for estimating the maximum risk to the population, a value important to a regulatory determination by a public health official. More recently, researchers have developed statistical distributions, and their standard deviations, of many of the important exposure variables. This allows a decision-maker to evaluate, for example, the 95% (two standard deviations) and 99% (three standard deviations) confidence intervals on the data, rather than forcing the use of an unrealistic maximum value. This has led to a significant reduction in many estimates of risk because the data show real exposures are almost always significantly lower than the maximum estimate. Considerable research has also been conducted to develop better lifestyle and activity patterns for humans. Rather than assuming that a person sits on his or her front porch continuously for 70 years, we can now portray with much greater certainty the time that humans spend at home, in transit, at work, during shopping, and at leisure. These distinctions are important because humans are typically exposed to different pollutants and at different concentrations during each of these different activities.

Much of this information resulted from important contributions by indoor air researchers who attempted to better define indoor air exposures and exposure patterns. The TEAM studies conducted by the EPA and others, described in Chapter 8, provided major new insights into total human exposures to pollutants, and they conclusively showed the importance of the indoor microenvironment in the assessment of total exposure. More recently, research is focusing on multipathway exposures and risks. In these efforts, the exposures and risks from all environmental pathways are being combined. In the past, regulators typically focused separately on each individual pathway. The legislation and the programmatic responsibilities in the regulatory agencies were usually separated for the different environmental media. In addition, scientific capabilities were not sufficiently advanced to consider the different media in combination. However, it was widely understood that many pollutants exist in more than one media and that humans can come into contact with multiple media. Not only are people exposed to many pollutants in this manner, but people often are also exposed to the same pollutant simultaneously from different media. This is leading to considerable research and, for some pollutants, a much better understanding of their total impacts on humans. Important research areas include studying bioaccumulation through the food chain, conducting particle deposition studies, and studying the chemical and physical changes in pollutants in the environment.

Risk characterization, the final step in the risk assessment process, brings together the relevant information from the hazard identification, dose–response, and exposure assessment work, and estimates (1) how likely the risk is to occur, and (2) what the consequences are if it does occur. Risk characterization is not an independent step that requires specific research. However, risk assessment guideline documents, in a sense, describe risk characterization as part of the risk assessment process, and risk assessment guidelines continue to undergo development, evaluation, and change. For example, the EPA published its first guidance on carcinogenic risk assessment in 1976 and revised it in 1986. A newly proposed version of the carcinogen risk assessment guidelines, released in 1996, was being publicly reviewed at the time this book was being written. At the same time, other EPA guidelines were in various stages of review and completion, including exposure assessment guidelines and ecological risk assessment guidelines.

B. Indoor Air Risk Assessment Research

Indoor air research is being conducted in the same areas as general risk assessment research, namely hazard identification, dose–response assessment, exposure assessment, and risk characterization. In addition, many indoor specific subjects are receiving close attention. Some of the more important areas of indoor air research related to risk assessment are described below.

An important indoor air research area is developing a better understanding of conditions that have come to be known as Sick Building Syndrome (SBS) and Multiple Chemical Sensitivity (MCS). Some buildings appear to be associated with a range of symptoms sufficiently consistent to be tagged as "sick" buildings. Chapter 3 describes research using chemosensory reactions recorded in conjunction with psychophysical or rating scale measures of sensory irritation to objectively evaluate the effects of volatile organic compounds, distinguish between olfactory and trigeminal components of sick building syndrome, and assess the reported hypersensitivity of multiple chemical sensitivity patients to chemicals. Chapter 3 also discusses research attempting to link VOC exposures to the development of sick building syndrome. Many of the VOCs detected indoors are neurotoxic, and clinical signs of VOC exposures can include headache, nausea, irritation of the eyes, mucous membranes, and the respiratory system, drowsiness, fatigue, general malaise, and asthmatic symptoms. Studies of the relation between exposure to indoor air VOCs and SBS to date show only sparse or inconsistent associations between observed VOC levels and health effects. Uncertain exposure assessment and symptom registration, as well as limitations within study designs, have been considered contributing factors. Some researchers note that factors other than chemical exposure may play a role in increased sensitivity in some individuals. These include comfort variables (i.e., heat and humidity), ventilation parameters, microbiological contaminations, and less common airborne pollutants typically ignored in indoor air studies. All of these issues point to the need for more research to confirm or repudiate the existence of SBS and MCS and, if confirmed, identify the root causes and ultimate solutions.

As noted in earlier chapters, public health officials must make decisions such that if there is error, it is on the side of public health protection. Because there was considerable uncertainty in early risk assessments, both outdoor and indoor, the assumptions made to fill the data gaps were usually conservative, meaning health protective. The potentially unrealistic outcome of this conservatism was recognized, but data were generally not available in early years to improve the process. Since then, considerable research has been undertaken to reduce the typically conservative default values applied when there is uncertainty. For example, better models are now available to predict more precisely the health effects associated with varying exposures. An example is the MKV model, discussed in Chapter 2, that includes consideration of cell turnover rates and other nongenotoxic events. In addition, total exposures to pollutants can now be estimated with much greater accuracy as a result of direct measurements in indoor environments and by better definition of population life-styles and exposure-producing habits; these improvements result from substantial indoor air research.

Recent research into the mechanisms of toxic effects has resulted in the development of new and useful procedures. For example, Chapter 3 describes test approaches utilizing chemosensory evoked potentials (CSEPs), visual evoked potentials (VEPs), and neurobehavioral changes to evaluate the effects of acute and chronic chemical exposure. Interestingly, numerous chemicals, including solvents, metals, and pesticides that are typical indoor air pollutants, were reported to alter VEPs in humans and animals. This may provide a useful extrapolation method of comparing toxic results in animal tests to toxic results in humans. CSEPs also appear to be useful because odors and sensory irritation of the eyes, nose, and throat provide early warning signs of potential toxic hazard. Neurobehavioral tests of sensorimotor and cognitive functions in children appear to be useful in assessing adverse effects of low-level chemical exposures.

A particularly controversial research area at the time this book was written involved understanding the effects of exposure to fine (equal to or less than 10 microns) particulate matter on mortality and illness. At the heart of the controversy are studies showing that mortality and illness increase with increasing exposures to fine particulate matter although no specific scientific mechanism had been proposed to explain the measured effect. Some scientists argue that other factors, as yet unmeasured, may be at play; others express concern that the test data are not being made available to the scientific community for further assessment and verification of the reported results. Notwithstanding the controversies, the EPA moved ahead under court order and promulgated (62 FR 38652, July 18, 1997), more restrictive standards for fine particulate matter. At the time this book was written, additional research was under way along various fronts. This research was taking on major significance because the revised standards have a potential, when implemented, for substantial economic impacts. Although the revised particulate matter standards apply only to outdoor exposures, the EPA actions and the research potentially affect the indoors. Most importantly, there are indoor sources of fine particulate matter; indeed, some scientists believe that the fact that people typically spend about 90% of their time indoors may be playing a significant role in the reported findings. This is leading to more research on the distributions and sources of indoor particulate matter and their relationship to outdoor levels.

One striking result of indoor air quality studies to date is the general lack of strong, definitive associations between exposure to indoor air pollutants and adverse health effects. This may result in part from the lack of properly designed epidemiological studies or the lack of appropriately sensitive test methods. More likely, it results from an array of problems including lack of data on the long-term effects of exposure to low concentrations of indoor air pollutants; questions about the relative role of indoor and outdoor air pollutants; potential confounding by tobacco smoking and chronic respiratory diseases; and the uncertain effects of exposure to biological contaminants. Research is under way in most of these areas and should provide useful results in the future.

Assessment of indoor air exposures has benefited from considerable research aimed at developing personal monitors and biological markers to measure more precisely human exposures to air pollutants. Personal monitors used in or near the breathing zone can be valuable tools for directly measuring a specific individual's exposure to a contaminant or group of contaminants. However, the technical challenges of designing nonintrusive instruments with sufficient sensitivity to the many different substances to which humans can be exposed indoors are considerable. This has led to significant ongoing research. Biological markers, discussed in Chapter 5, are valuable means for confirming previous exposures to specific substances. While these have been used primarily in limited areas (e.g., nicotine and carbon monoxide exposures), they show great promise for broader use. Researchers are studying ways to use these tests in a less invasive manner by less highly trained personnel, and the biological variations in humans that can result in test differences are being explored.

One interesting area of research is the development of more appropriate survey tools to gather personal information from subjects exposed in indoor air quality studies. Questionnaires are frequently difficult to interpret because of the vagaries and uncertainties of human response. Considerable research is under way to develop more precise and more easily used tools to improve the quality and quantity of the personal information gathered.

One troubling issue is the significant national growth in cases of bronchial asthma since the early 1970s; the beginnings of this growth roughly coincided with the oil embargo and the improvement in indoor air management to reduce energy consumption. Many researchers are studying this issue with a focus on the possible effects of exposure to dust mite allergens. Other researchers believe that they have correlated further increases in urban asthma cases with the increased use of methyl-*tert*-butyl ether (MTBE) in gasoline, an additive used to increase oxygen content and thus reduce emissions of harmful pollutants.

VOCs represent a substantial category for research because there are literally thousands, if not hundreds of thousands, of VOCs to which humans can be exposed. At the time this book was written, over 1,000 specific VOCs were regulated as air pollutants by federal, state, or local air agencies. Considerable research also was under way to develop more accurate monitors to measure these substances and more accurate mathematical models to interpret the results.

IV. DIRECTIONS IN RISK ASSESSMENT RESEARCH

The National Research Council (NRC 1994) identified six important themes that cut across the various stages of risk assessment. Each theme is described below and areas upon which research should focus are identified.

Default Options: Is there a set of clear and consistent principles for choosing and departing from default options? The NRC recommended the continued use of default options as a reasonable way to cope with uncertainty. However, the use of each default option should be clearly identified, the scientific and policy basis for the option fully explained, and criteria for departure given greater formality.

Methods and Models: Are the methods and models used in risk assessment consistent with current scientific information? The NRC recommended a number of actions relating to methods and models, including improvements in emission characterization, exposure assessment models and databases, and toxicity assessment methods and models.

Data Needs: Are sufficient data available to generate risk assessments that protect the public health and are scientifically plausible? The NRC made a number of recommendations. For example, the 189 hazardous air pollutants listed in the 1990 Clean Air Act Amendments should be screened for priorities for assessment of health risks, and a database of exposure information on these pollutants should be developed. An iterative approach to gathering and evaluating data in both screening and full risk assessments should also be defined and developed. Finally, data management systems must be improved to ensure that the quality and quantity of risk assessment data are sufficiently accessible and routinely updated.

Uncertainty: Is the inevitable uncertainty in risk assessment sufficiently accounted for in the consideration, description, and decisions being made using the risk information? The NRC again made a number of recommendations. For example, single point estimates should not necessarily be abandoned, but these numbers must be based on careful consideration of both the estimate of risk and its uncertainty. Also, uncertainties should be made explicitly, presented as accurately and fully as feasible, and presented quantitatively to the extent feasible.

Variability: Is the extensive variation among individuals in their exposures to toxic substances and in their susceptibilities to cancer and other health effects sufficiently considered? NRC recommendations include the following: distributions of exposure values should be developed to the extent possible based on available measurements, modeling results, or both; the EPA, the National Institutes of Health, and other federal agencies should sponsor molecular, epidemiologic, and other research on the extent of interindividual variability in factors that affect susceptibility and cancer; and separate risk estimates should be determined for adults and children where there is reason to believe that the risks may be related to age.

Aggregation: Is the possibility of interactions among pollutants in their effects on human health as well as the consideration of multiple pathways and multiple adverse effects sufficiently considered? The NRC recommended research in several areas relevant to aggregation. For example, multiple routes of exposure and multiple end points should be considered more frequently and to the extent that data are available for aggregating cancer risks.

The EPA Science Advisory Board (SAB) Indoor Air Quality and Total Human Exposure Committee also reported to the EPA Administrator in 1995 (EPA 1995) on the broader issue of total human exposure but with an emphasis on the contribution of indoor air quality. Based on their study, the Committee made five specific recommendations to the EPA that have indoor air implications.

1. Develop a mechanism to support the research, validation, and application of: (a) more sensitive and specific microsensors, biomarkers, and other monitoring technologies and approaches for measuring exposures, and (b) validated data on associated exposure determinants, including demographic characteristics, time-activity patterns, locations of activities, and behavioral and life-style factors.
2. Establish a mechanism to develop, validate with field data, and iteratively improve models that integrate: (a) measurements of total exposure and their determinants, (b) a better knowledge of exposure distributions across different populations, and (c) the most current understanding of exposure–dose relationships.
3. Develop, in cooperation with other agencies and stakeholders, a robust database that reflects the status and trends in national exposure to environmental contaminants.
4. Develop sustained mechanisms and incentives to ensure a greater degree of interdisciplinary collaboration in exposure assessment and, by extension, in risk assessment and risk management studies.
5. Take advantage of improving capabilities in exposure assessment technology, electronic handling of data, and electronic communications to establish and disseminate early warnings of emerging environmental stressors.

More specifically, the Committee identified three examples of new sensor technologies with considerable potential application to air pollutant exposure assessment. Highly sensitive ultrasonic flexural plate wave (FPW) devices are being developed for *in situ*, real-time analyses of particles and VOCs in indoor and outdoor environments. These sensors can be batch fabricated using well-developed and inexpensive silicon technology and interfaced with microprocessors that record and analyze the sensed measurements. In addition, excimer laser fragmentation/fluorescence spectroscopy (ELFFS) is being used to detect metals and organics in the part per billion range. The method is nonintrusive, fast, and can selectively detect and quantify many substances. Lastly, computer tomography/Fourier transform infrared spectrometry is an emerging technology that can characterize spatial distributions and movements of air pollutants in three dimensions in indoor and outdoor environments. The technology is expected to be commercially available by the turn of the century.

The Committee also identified several important areas of indoor air quality in need of research. First, federal, state, and local agencies must make fundamental changes in their approaches to environmental monitoring. The commonly used approach of sampling single contaminants, single media, and single pathways, with no clear relationship to the time-activities of those exposed, will not be adequate for addressing future needs. Second, environmental monitoring efforts are typically conducted for regulatory purposes, with the regulations representing a patchwork of perceived needs and partial solutions; these efforts must be broadened to assess complex contaminant mixtures and to relate the exposures to dose and, ultimately, to the endpoints of concern. Third, personal inhalation monitors need to be improved to provide substantially more information, including concurrent measurements of breathing rates and exercise patterns, as well as the accompanying composition of the individual's exhaled air. Finally, biomarkers can serve as indicators of exposure, dose, susceptibility, preclinical disease, and biological injury and disease processes. While proven to be valuable in research, they have yet to achieve success as practical indicators of population susceptibility, exposure, or response. Expanded research is needed in the development of biomarkers and in considering the ethical issues inherent in applying biomarkers; these issues include false alarms and the needless stress for individuals warned about the presence of uncertain signals.

BIBLIOGRAPHY

Center for Indoor Air Research (CIAR). 1996. *1996–1997 Research Agenda, Request for Applications*, Linthicum, MD.

Chemical Manufacturers Association (CMA). 1988. *Chemicals in the Community: Methods to Evaluate Airborne Chemical Levels*, April, 1988, Chemical Manufacturers Association: Washington, DC.

Department of Health and Human Services (DHHS). 1985. Risk Assessment and Risk Management of Toxic Substances, A Report to the Secretary, Department of Health and Human Services from the DHHS Committee to Coordinate Environmental and Related Programs (CCERP), April, 1985.

Duan, N. 1985. Application of the Microenvironment Monitoring Approach to Assess Human Exposure to Carbon Monoxide, R-3222-EPA, prepared for the U.S. Environmental Protection Agency by Rand, Santa Monica, CA.

Environ. 1986. *Elements of Toxicology and Chemical Risk Assessment, A Handbook for Nonscientists, Attorneys and Decision Makers,* Environ Corporation: Washington, DC.

Environmental Protection Agency (EPA). 1979. National emission standards for identifying, assessing and regulating airborne substances posing a risk of cancer, *44 FR 58642,* October 10, 1979

Environmental Protection Agency (EPA). 1981. Policy and Procedures for Identifying, Assessing and Regulating Airborne Substances Posing a Risk of Cancer, Draft, 6/22/81, Pollutant Assessment Branch, Strategies and Air Standards Division, Office of Air Quality Planning and Standards, U.S. Environmental Protection Agency.

Environmental Protection Agency (EPA). 1985. Bibliographic Series, Report No. EPA/IMSD-85-002, Information Services and Library, U.S. Environmental Protection Agency, Washington, DC.

Environmental Protection Agency (EPA). 1986a. Part II. Guidelines for carcinogen risk assessment, *51 FR 33992,* September 24, 1986.

Environmental Protection Agency (EPA). 1986b. Part III. Guidelines for mutagenicity risk assessment, *51 FR 34006,* September 24, 1986.

Environmental Protection Agency (EPA). 1986c. Part IV. Guidelines for health risk assessment of chemical mixtures, *51 FR 34014,* September 24, 1986.

Environmental Protection Agency (EPA). 1986d. Part V. Guidelines for health assessment of suspect developmental toxicants, *51 FR 34028,* September 24, 1986.

Environmental Protection Agency (EPA). 1986e. Part VI. Guidelines for exposure assessment, *51 FR 34042,* September 24, 1986.

Environmental Protection Agency (EPA). 1989. Report to Congress on Indoor Air Quality, FPA/400/1-89/001A, Office of Air and Radiation and Office of Research and Development, U.S. Environmental Protection Agency, August 1989.

Environmental Protection Agency (EPA). 1990. *Managing Asbestos in Place, A Building Owner's Guide to Operations and Maintenance Programs for Asbestos-Containing Materials, 20T-2003, Pesticides and Toxic Substances,* U.S. Environmental Protection Agency.

Environmental Protection Agency (EPA). 1991. Testimony of Deputy Administrator F. Henry Habicht, Jr. before the Subcommittee on Health and the Environment, U.S. House of Representatives, April 10, 1991.

Environmental Protection Agency (EPA). 1995. An SAB Report: Human Exposure Assessment — A Guide to Risk Ranking, Risk Reduction, and Research Planning. Report No. EPA-SAB-IAQC-95-005, Science Advisory Board, U.S. Environmental Protection Agency, March 1995.

Foster, S.A., Chrostowski, P.C. 1986. Integrated household exposure model for use of tap water contaminated with volatile organic chemicals, for presentation at the 79th annual meeting of APCA, Minneapolis, MN, June 22–27, 1986.

Foster, S.A., Chrostowski, P.C. 1987. Inhalation exposures to volatile organic contaminants in the shower, for presentation at the 80th annual meeting of APCA, New York, NY, June 21–26, 1987.

Lippman, M. 1987. Feasibility of field studies of multipollutant interactions, for presentation at the 80th annual meeting of APCA, New York, NY, June 21–26, 1987.

National Reseach Council (NRC). 1983. *Risk Assessment in the Federal Government: Managing the Process*, Committee on the Institutional Means for Assessment of Risks to Public Health, Commission on Life Sciences, National Research Council: Washington, D.C.

National Research Council (NRC). 1994. *Science and Judgment in Risk Assessment*, prepared by the Committee on Risk Assessment of Hazardous Air Pollutants, Commission on Life Sciences, National Academy Press: Washington, DC, 1994.

Stolwijk, J.A.J. 1987. Multipollutant indoor exposures and health responses: Epidemiological approaches, for presentation at the 80th annual meeting of APCA, New York, NY, June 21–26, 1987.

Index

A

Absorbed dose, 6
Acceptable daily intake (ADI), 47
Acceptable risk, 6
Accuracy, 6
1-Acetoxy-2-ethoxyethane, 186
ACGIH, see American Conference of
    Governmental Industrial Hygienists
ACOP, see Approved Codes of Practice
Acrolein, 63
ACS, see American Cancer Society
Activated charcoal, 168
ACTS, see Advisory Committee on Toxic
    Substances
Acute effect, 6
Acute exposure, 6
Acute risks, 2
ADI, see Acceptable daily intake
Advisory Committee on Toxic Substances
    (ACTS), 17
Aerodynamic Particle Sizer (APS), 167
Age-at-exposure model, 91
Agency for Toxic Substances and Disease
    Registry (ATSDR), 116
Aggregation, 238
AHERA, see Asbestos Hazard Emergency
    Response Act
Air
  exchange, 172
    measurements, 185
    rate, 178, 193
  freshener, 189
  pollution, 224
    debate, 13
    mortality from short-term exposure to,
      217

Airborne contagious diseases, 64
Airflow modeling, 92
ALA, see American Lung Association
Alcohol consumption, association between
    oral cancer and, 78
Aldrin, 163, 171
Alkylcyclohexanes, 186
Allergens, measuring indoor, 59
Alzheimer's disease, 66
Ambient air monitoring data, 226
American Cancer Society (ACS), 219
American Conference of Governmental
    Industrial Hygienists (ACGIH), 5,
    216, 232
American Lung Association (ALA),
    214
American Society of Heating, Refrigerating,
    and Air-Conditioning Engineers
    (ASHRAE), 232
American Society for Testing and Materials
    (ASTM), 29
Animal
  -bioassay data, 54, 55
  carcinogenicity, 149
  -to-human extrapolation, 145
  studies, 40, 67
Antagonism, 6
Applied dose, 6, 100
Approved Codes of Practice (ACOP),
    16
APS, see Aerodynamic Particle Sizer
AREAL, see Atmospheric Research and
    Exposure Assessment Laboratory
Armitage–Doll approximation, 85
Aromatic hydrocarbons, 117
Aromatics, 186
Asbestos, 22, 103